A Practical Guide
to Supramolecular Chemistry

A Practical Guide to Supramolecular Chemistry

Peter J. Cragg

School of Pharmacy and Biomolecular Sciences,
University of Brighton

John Wiley & Sons, Ltd

Other Wiley Editorial Offices

John Wiley & Sons Inc., 111 River Street, Hoboken, NJ 07030, USA

Jossey-Bass, 989 Market Street, San Francisco, CA 94103-1741, USA

Wiley-VCH Verlag GmbH, Boschstr. 12, D-69469 Weinheim, Germany

John Wiley & Sons Australia Ltd, 42 McDougall Street, Milton, Queensland 4064, Australia

John Wiley & Sons (Asia) Pte Ltd, 2 Clementi Loop #02-01, Jin Xing Distripark, Singapore 129809

John Wiley & Sons Canada Ltd, 22 Worcester Road, Etobicoke, Ontario, Canada M9W 1L1

Wiley also publishes its books in a variety of electronic formats. Some content that appears in print may not
be available in electronic books.

Library of Congress Cataloging-in-Publication Data

British Library Cataloguing in Publication Data

A catalogue record for this book is available from the British Library

ISBN-13 978-0-470-86653-5 (HB)   978-0-470-86654-3(PB)
ISBN-10 0-470-86653-5 (HB)   0-470-86654-3 (PB)

For Alex and James

Contents

Preface

Supramolecular chemistry – the branch of chemistry associated with the formation of complex multimolecular entities from relatively simple molecular components – has been a major research theme over the past four decades. Researchers were provided with a unifying vision of the field following the award of the 1987 Nobel Prize in chemistry to three of its pioneers: Jean-Marie Lehn, Donald Cram and Charles Pedersen. Subsequently there have been a number of excellent texts on the subject ranging from undergraduate primers and books on supramolecular design to comprehensive works and encyclopaedias. To date the theoretical aspects of this fascinating area of science have been well served by texts at all levels. As far as laboratory-based supramolecular chemistry is concerned there are some edited works in which research teams have presented the syntheses of compounds that they pioneered. By and large the syntheses are extended forays into specific groups of compounds which, while they are invaluable to the experienced researcher, may seem a little daunting to the neophyte. Isolated examples of other synthetic methods are to be found in a number of undergraduate practical guides; however, they are rarely placed in a supramolecular context. Finally, one or two classic syntheses of compounds, including crown ethers and calixarenes, are to be found among the pages of the collective volumes of *Organic Syntheses*.

This book is intended to take the historical and theoretical background of supramolecular chemistry into the laboratory and to be used as an entry level synthetic guide for those with little or no prior experience in supramolecular chemistry yet who wish to incorporate aspects of it in their own research or teaching. It incorporates practical syntheses designed so that chemists who are not necessarily supramolecular specialists can prepare archetypal compounds used in supramolecular chemistry – crown ethers, podands, resorcinarenes, calixarenes and the like – using straightforward experimental procedures. All syntheses are simple enough to be undertaken using equipment available in most university and college chemistry laboratories. As a result the procedures are not necessarily those with the highest yields but those that have been 'tried and tested' as the most reproducible. Techniques commonly used in the prediction and analysis of supramolecular phenomena are also discussed with some examples described in detail. Compounds that are prepared from directions in the text can then be used to illustrate particular supramolecular phenomena such as clathrate formation or may be developed further as part of the researcher's own work. Experimental procedures for representative ligands in each class are preceded by brief outlines of the historical and current interest in them. Given the vast scope of contemporary supramolecular chemistry this book cannot hope to cover every class of compound or experimental technique of interest to the community. Many compounds were prepared during the research for this book but failed to be included either because the methods required specialist equipment, not always to hand in an undergraduate laboratory, or because of complex work-up procedures. To aid in the identification of products, the compounds described are accompanied by approximate yield, melting point (where appropriate), infrared and ^{1}H nuclear magnetic resonance data. Note that most values have been rounded up or down so that infrared data, for example, are reported generally to the nearest five wavenumbers. As trace impurities, often solvents, can have a significant effect on spectral and melting point values, the data reported should be used as a guide.

In some cases the synthesis given is based closely on one available in the literature, albeit sometimes from a quite obscure paper, and where this is the case the relevant authors and papers are clearly cited. These references should be given in any published works where the compounds appear.

Acknowledgements

Firstly I must acknowledge a great debt to the many supramolecular chemists past and present who have pioneered or developed many of the synthetic routes described herein. As will be clear from an examination of the cited literature, there are several prolific individuals and their research groups without whom some exciting supramolecular avenues would have remained unexplored. There are, of course, many others who have contributed to the field and have helped to make supramolecular chemistry the vibrant discipline it is today. Alongside this must go thanks to members of my own research group who have been joining me on my adventures in supramolecular chemistry for the past decade. Thanks to Flavia Fucassi who ran many of the spectra and repeated some of my experiments to prove that they worked. Thanks also to Andy Slade and Rachael Ballard at Wiley who twisted my arm to write this book and then left me in peace for two years while I got on with the job. The experimental work was carried out in the School of Pharmacy and Biomolecular Sciences at the University of Brighton. All the *in silico* simulations and computer graphics were generated by Spartan '03 software (Wavefunction, Inc., Irvine, CA). Finally, and most importantly, thanks to Margaret for all her support – ever since we met in that supramolecular lab in Tuscaloosa.

Peter J. Cragg

Introduction

So What is 'Supramolecular' Chemistry?

The term 'supramolecular' has been used for many years in the context of complex biological structures however the use of this label to describe molecular scale interactions dates back to Lehn's Nobel Prize address in 1987 where it was defined as 'chemistry beyond the molecule' [1]. Since then the number of papers citing 'supramolecular' as a keyword has grown almost exponentially it has found its way into the titles of textbooks, specialist series and periodicals. Other terms that are used interchangeably with supramolecular chemistry include 'inclusion phenomena', 'host–guest chemistry' or 'molecular recognition'. The former is a broadly applied synonym whereas the latter two have their origins in the 'lock and key' mechanism of biological catalysis proposed by Emil Fischer in 1894 [2]. While these are useful key words for the subject they do not directly answer the question. What exactly makes a particular aspect of chemistry 'supramolecular'?

Technically almost any reaction can be classed as operating on a scale that is 'beyond the molecule' whether it is the catalysis of polypropylene, the solidification of cement or gene transcription. What makes the concept of supramolecular chemistry attractive is that there is an implicit act of design in the use of existing molecules, or in the synthesis of new ones, to prepare molecular assemblies that have particularly desirable qualities. Such qualities may include detection of biologically important species in a hospital environment, monitoring agrochemicals in watercourses, controlled release of pharmaceuticals into a patient or even the ability for a supramolecular interaction to act as a 'bit' of information on a chemical computer chip.

A good example of the impact of supramolecular research is in the design and synthesis of molecules that respond to a particular analyte. To successfully target a specific chemical species it is necessary to match its size and binding preferences

A Practical Guide to Supramolecular Chemistry   Peter J. Cragg
© 2005 John Wiley & Sons, Ltd

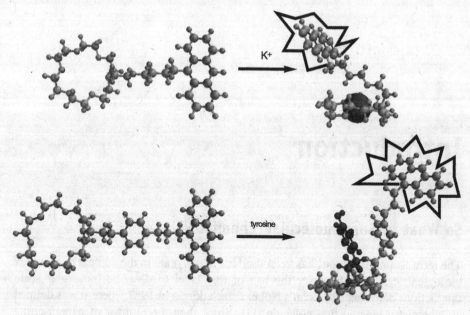

Figure 1 Potential supramolecular sensors for potassium (top) and tyrosine (bottom)

to a receptor molecule that will bind the analyte of interest in preference to all others. The requirements for a potassium sensor (Figure 1) will therefore be very different from one for a transition metal such as cobalt or even another alkali metal like sodium, potassium's next-door neighbour in the Periodic Table. Likewise, a receptor for tyrosine will have different requirements than one for arginine even though they are both amino acids. Matching the receptor to the analyte may be considered to be an exercise in supramolecular chemistry. Why is this so? The field of supramolecular chemistry is now quite mature and, as a result, there is a remarkable amount of information about molecular hosts and their affinities for different guests available in the scientific literature. Thus a scientist well versed in the literature will be able to choose the best complementary binding motif for a specific analyte. [18]Crown-6 is a good receptor for the protonated N-terminus of an amino acid and also for the potassium cation. It could therefore form part of a receptor for both of the above examples. To make the receptors more specific further functionality must be incorporated. In the case of potassium the attachment of a short polyether side chain improves binding and can be achieved using an aza[18]crown-6 derivative. Conversely, the phenolic group in tyrosine will engage in π–π interactions readily so incorporation of a substituent containing an aromatic group would give greater specificity for tyrosine. The next requirement is that the binding of the analyte – the 'host–guest' interaction – to the receptor must trigger an observable response through a reporter molecule. The nature of the response will depend upon the analytical method used to detect the signal but will typically be colorimetric, electrochemical or fluorescent. An appropriate signalling mechanism may be designed to incorporate, for example, changes in conformation or molecular orbital energies that in turn affect the reporter molecule's electronic or spectroscopic signature upon analyte binding. Finally, the receptor must be linked to the reporter molecule such that the effect of binding directly affects the latter without compromising the observable response. Sensors for the two analytes above could both be based on aza[18]crown-6 and contain a terminal anthracene group as a fluorescent reporter but differ in the link between them. An ether link may suffice for one and an aromatic spacer for the other, as illustrated above.

These generalizations illustrate some of the design issues that emerge when even a simple response is required. There is a further complication to be considered if the sensor is intended to work in an aqueous environment, for example to detect the target metal in river water samples or blood, as it ought to be water soluble but unaffected by the effect of solvation. The sensor is therefore responding to fundamental 'supramolecular' phenomena: highly specific analyte–receptor binding inducing a change in an observable aspect of the parent molecule. It is an approach that necessitates a high degree of design prior to synthesis and it is this aspect of molecular level design which sets the supramolecular sensor apart from other methods of analysis such as the precipitation of a coloured complex salt upon addition of a specific reagent. Given the ever-increasing levels of sophistication in optics, electronics and microscopy it is conceivable that a carefully designed sensor based on supramolecular principles could soon be used to detect single molecules, something that is unlikely to be true of more traditional complexation methods.

Supramolecular Synthons, Assembly and Phenomena

Given the somewhat eclectic nature of the compounds and methods included in this book, the author does not expected it to be read from cover to cover, rather that researchers will wish to dip in to various sections in search of a particular synthesis or technique. As a starting point to investigate any aspect of supramolecular chemistry it is likely that a researcher will need to acquire one or more of the 'building blocks' described in the growing numbers of scientific papers published in the field. By and large the most important group of compounds encountered are large cyclic molecules, or macrocycles, that incorporate potential binding sites for guest molecules. These macrocycles have different degrees of cavity pre-organization, that is the relative positions of the binding sites may be rigidly fixed in space to direct lone pairs of electrons into the central cavity, or they may have greater degrees of conformational freedom. Examples of the former include cryptands, porphyrins and phthalocyanines whereas the latter are typified by crown ethers. As the main emphasis in this book is to use these compounds to prepare 'supramolecules' with specifically designed functions they will be referred to as supramolecular synthons by analogy to the use of precursor molecular synthons to prepare covalently linked molecules. The level of preorganization in the examples described increases in complexity throughout the book, from flexible polyether derivatives through podands to crown ethers, calixarenes and cryptands. The preparation of a variety of derivatives is described for almost every class of synthon. There is a bias towards crown ethers and expanded calix[3]arenes, as they have been the mainstay of the author's research for almost two decades, though a diversity of structural and binding motifs are represented by the examples described. The classes of compounds commonly used by supramolecular chemists are illustrated below.

Supramolecular assembly, the formation of supramolecules from supramolecular synthons (Figure 2), is discussed with reference to methods by which the

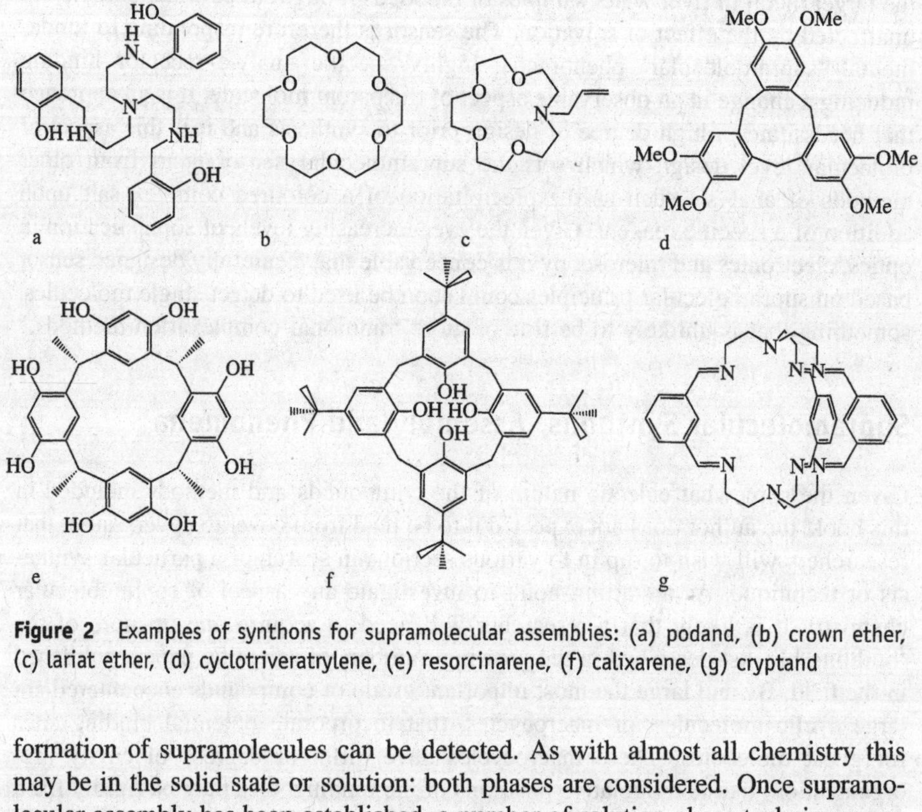

Figure 2 Examples of synthons for supramolecular assemblies: (a) podand, (b) crown ether, (c) lariat ether, (d) cyclotriveratrylene, (e) resorcinarene, (f) calixarene, (g) cryptand

formation of supramolecules can be detected. As with almost all chemistry this may be in the solid state or solution: both phases are considered. Once supramolecular assembly has been established, a number of solution phase techniques can be used to generate data on the stoichiometry of the species involved and the affinities of host molecules for guests. To illustrate these, the methods to determine host–guest binding are discussed and one example given. The importance of predictive computational methods is considered so the main techniques available to non-specialist researchers are described to help determine the most appropriate computational approach to a supramolecular problem of interest. Given the complexity of supramolecular assemblies, *in silico* chemistry is fast becoming a valuable method by which they can be modelled in an attempt to understand structures and mechanistic details; however, it is important to understand limitations to the technique.

Supramolecular phenomena (specifically clathrate formation), general inclusion phenomena and self-assembly, are described and some examples are given using synthons described in the text. Given that initial interest in many supramolecule-forming compounds relied on their similarities to natural products, it is fascinating to report on recent advances in the use of DNA as a molecular scaffold or calixarene self-assembly that mimics icosahedral viral coatings. Throughout the book it will be seen that the natural world remains an inspiration behind some of the major discoveries in supramolecular chemistry.

Hazards of Supramolecular Chemistry

Synthetic chemistry is fraught with danger for the inexperienced or indecisive researcher. Supramolecular chemistry is no different. Every compound presented in this book has been prepared personally by the author who, in addition to an excellent undergraduate training, has had the privilege to learn from some outstanding supramolecular chemists since embarking on a PhD in the field in 1986. Thus all of the experiments described have been undertaken without incident. Any particular hazards are clearly identified and may be cross referenced with a complete appendix of all reagents and solvents used. Unfortunately there are always unforeseen problems. Typically these result from carelessness and inattention to detail. An example of this is the presence of a small star crack in a round-bottomed flask that has been stored poorly. Once the flask is subjected to reduced pressure, as occurs in rotary evaporation of excess solvent, it may well implode with disastrous results, particularly if the experimentalist is not wearing suitable protective clothing. Another potentially hazardous situation is one in which flasks, beakers or bottles containing reagents are left unlabelled. An almost infinite variation of colourless and odourless, or nearly odourless, liquids can be imagined, from solutions of acids and bases to volatile, flammable solvents and aqueous solutions of highly toxic reagents. The consequences of the experimentalist, or more often a colleague in the same laboratory, mistaking the contents of the vessel could be catastrophic. It is of utmost importance that, however straightforward a particular experimental procedure appears, full consideration is taken of the inherent and potential risks. As many of the syntheses included here are suitable for undergraduates, either in a fully supervised laboratory session or unsupervised as part of an undergraduate research project, it is important that supervisors are also aware of any risks.

Safety Considerations

The following list is not exhaustive and other points could be made. It is intended as a basic set of guidelines to be considered prior to undertaking any of the syntheses described in this book. The overarching consideration must be to conform to all safety regulations as set out by the relevant safety officers or committee in the researcher's work place.

- Always plan the experiment in advance and note any unfamiliar methods or potentially hazardous procedures. If you are supervising staff or students who will carry out the experiment make sure that you explain any unfamiliar procedures clearly and concisely.

- Always wear suitable protective clothing. A laboratory coat and safety glasses are a minimum; protective gloves are also advisable.

- Conform to any additional local safety regulations.

- Be aware of manufacturer's safety data sheet information and COSHH/OSHA details for chemicals that will be used and, if possible, for those that will be prepared during the experiment. If no such data exist for the products assume that they will have the most hazardous properties of the reactants.

- Carefully check for incompatible reagents or methods such as heating flammable solvents with a naked flame.

- Ensure that the work area within the laboratory is safe and that any electrical equipment has been inspected recently.

- Work where possible in a fume hood.

- Carefully assemble all equipment, glassware, reagents and solvents prior to starting the experiment.

- Always ensure that equipment can be switched off safely and glassware taken apart quickly in the event of an emergency.

- Know where to find safety equipment such as fire blankets, extinguishers and eye rinsing facilities and how to contact staff trained in first aid.

- Dispose of waste solvents, solutions and precipitates using only the authorized procedures of your institution or workplace.

- Dispose of glass pipettes and any broken glassware using only the designated 'sharps' bins.

- Label all solutions transferred into beakers for the duration of the experiment and all products resulting from the reactions.

- Store all products carefully, bearing in mind their sensitivities to heat, light, air or moisture.

All the experiments described in the text can be undertaken without incident if directions are conscientiously followed by adequately trained researchers or supervised students.

[1] Supramolecular chemistry – scope and perspectives. Molecules, supermolecules, and molecular devices (Nobel lecture), J.-M. Lehn, *Angew. Chem. Int. Ed. Engl.*, 1988, **27**, 90.
[2] Einflus der Configuration auf die Wirkung der Enzyme, E. Fischer, *Ber. Dtsch., Chem. Ges.*, 1894, **27**, 2985.

A supramolecular bibliography

The following resources should be of interest to those wanting to know more about aspects of supramolecular chemistry.

Introductory texts
Supramolecular Chemistry (Oxford Chemistry Primers), P. D. Beer, P. A. Gale, D. K. Smith, Oxford University Press, Oxford, 1999.
Supramolecular Chemistry, J. W. Steed and J. L. Atwood, John Wiley & Sons, Ltd., Chichester, 2000.
Supramolecular Chemistry: Concepts and Perspectives, J.-M. Lehn, Wiley-VCH, Weinheim, 1995.
Supramolecular Chemistry: An Introduction, F. Vögtle, John Wiley & Sons Ltd., Chichester, 1993.

Specialized texts
Macrocycle Synthesis: A Practical Approach (Practical Approach in Chemistry Series), D. Parker (ed.), Oxford University Press, Oxford, 1996.
Principles and Methods in Supramolecular Chemistry, H. J. Schneider and A. Yatsimirsky, John Wiley & Sons Ltd., Chichester, 1999.
Supramolecular Chemistry of Anions, A. Bianchi, K. Bowman-James and E. Garcia-Espana (eds), Wiley-VCH, Weinheim, 1997.

Modified Cyclodextrins: Scaffolds and Templates for Supramolecular Chemistry, C. J. Easton and S. F. Lincoln, Imperial College Press, London, 1999.

Template Synthesis of Macrocyclic Compounds, N. V. Gerbeleu, V. B. Arion and J. Burgess (eds), Wiley-VCH, Weinheim, 1999.

Supramolecular Organometallic Chemistry, F. T. Edelmann and I. Haiduc, Wiley-VCH, Weinheim, 1999.

Analytical Chemistry of Macrocyclic and Supramolecular Compounds, S. M. Khopkar, Springer-Verlag, Berlin, 2002.

Book series and reference works

Monographs in Supramolecular Chemistry, Royal Society of Chemistry, London.

Perspectives in Supramolecular Chemistry, John Wiley & Sons Ltd., Chichester.

Advances in Supramolecular Chemistry, G. W. Gokel (ed.), JAI Press.

Organic Syntheses, Collective Volumes 1–10, J. P. Freeman (ed.), John Wiley & Sons Ltd., Chichester, 2004.

Encyclopedia of Supramolecular Chemistry, J. W. Steed and J. L. Atwood (eds), Marcel Dekker, New York, 2004.

Comprehensive Supramolecular Chemistry (Volumes 1–11), J. L. Atwood, J.-M. Lehn, J. E. D. Davies, D. D. MacNicol and F. Vögtle (eds), Pergamon, Oxford, 1996.

Journals dedicated to supramolecular chemistry

Supramolecular Chemistry, Taylor & Francis (Publisher), London.

Journal of Inclusion Phenomena and Macrocyclic Chemistry, Springer Science + Business Media B. V. (Publisher), London.

1

Linear Components for Supramolecular Networks

1.1 Flexible Components

The origins of supramolecular chemistry are intimately linked to Pedersen's pioneering work on crown ethers. His fortuitous discovery of dibenzo[18]crown-6 in 1967 occurred when he was attempting to prepare a flexible phenol-terminated polyether which was expected to bind to the vanadyl ion and thus support its catalytic activity [1]. Although the importance of the new class of cyclic polyethers was recognized across the globe, researchers were also alerted to Pedersen's original goal: the potential for acyclic polyethers to act as metal-binding ligands. Starting points for many of the experiments were the widely available polyethylene glycols that could be modified readily. Foremost among those preparing functionalized polyethers were Vögtle and Weber who coined the term 'podands' for their compounds [2].

Polyethers are well known for their abilities to wrap around metal ions, particularly those in groups 1 and 2 and the lanthanides (Figure 1.1), but the introduction of coordinating termini in concert with variable lengths of the polyether tether gave greater specificity to the ligands [3,4]. More recently, polyethers have been used to prepare coordination networks through the incorporation of transition metals. The inherent flexibility of the ether link coupled to the preferred geometry of certain metals gives rise to some very interesting nanoscale complexes, many of which have a helical motif [5].

Two methods to prepare polyether-based podands are described here. The first is a direct modification of polyethylene glycols and comes originally from the work of Piepers and Kellogg [6], later modified by Hosseini [7]. A simple reaction between hexaethylene glycol and isonicotinyl chloride, as shown in Figure 1.2, results in the formation of podand **1** in good yield. When treated with silver salts

A Practical Guide to Supramolecular Chemistry Peter J. Cragg
© 2005 John Wiley & Sons, Ltd

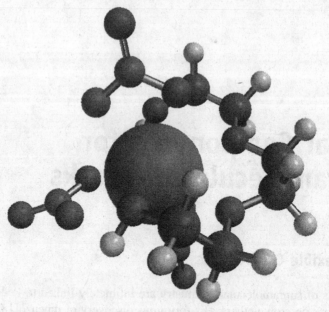

Figure 1.1 A triethylene glycol complex of europium(III) nitrate

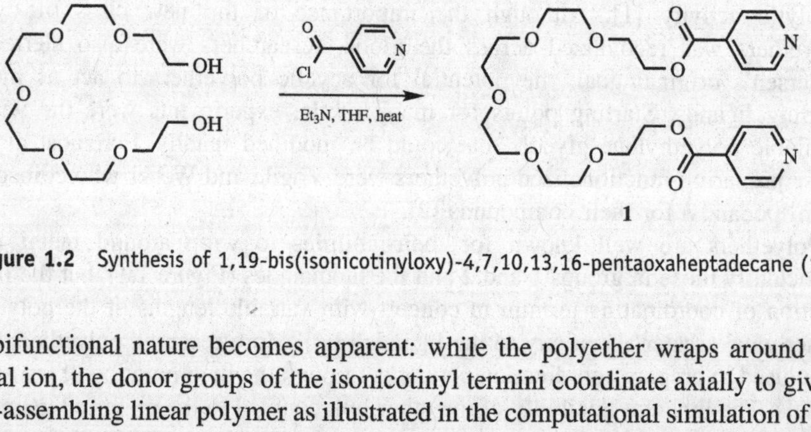

Figure 1.2 Synthesis of 1,19-bis(isonicotinyloxy)-4,7,10,13,16-pentaoxaheptadecane (**1**)

its bifunctional nature becomes apparent: while the polyether wraps around the metal ion, the donor groups of the isonicotinyl termini coordinate axially to give a self-assembling linear polymer as illustrated in the computational simulation of the X-ray structure (Figure 1.3). As shown by later work by the Hosseini group, many variations of the synthesis can be envisaged [8]. This method gives a yield that is in close agreement with the published 86 per cent.

The second method can be used to synthesize Vögtle-type podands [9,10] in two high-yielding steps from polyethylene glycol ditosylates. The ditosylate derivatives, which are also precursors of cyclic crown ethers, azacrown ethers and lariat ethers, can be prepared with pyridine as the base [11]. However, they are synthesized more effectively in higher yields using aqueous sodium hydroxide and tetrahydrofuran (Figure 1.4) [12]. These conditions also remove the necessity to

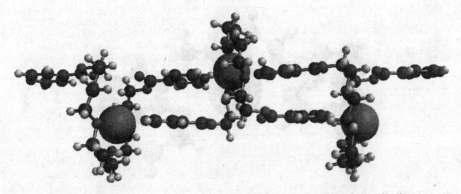

Figure 1.3 Part of the infinite linear cationic chain formed by **1** with silver

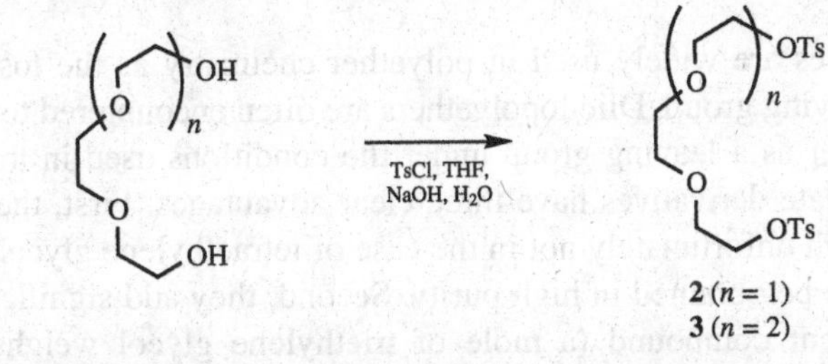

2 (*n* = 1)
3 (*n* = 2)

TsCl, THF,
NaOH, H₂O

Figure 1.4 Syntheses of triethyleneglycol ditosylate (**2**) and tetraethyleneglycol ditosylate (**3**)

work with large amounts of an unpleasant solvent. Furthermore it avoids problems inherent in the removal of excess pyridine from the reaction. Once the ditosylates have been prepared they can be treated with simple metal salts such as sodium 8-hydroxyquinolinate, as shown in Figure 1.5, to introduce useful termini. Although the products have high conformational mobility, the termini converge on alkali metals while the oxygen donors in the polyether backbone wrap around the metals as shown in crystal structures of related compounds and illustrated in Figure 1.6 [13,14].

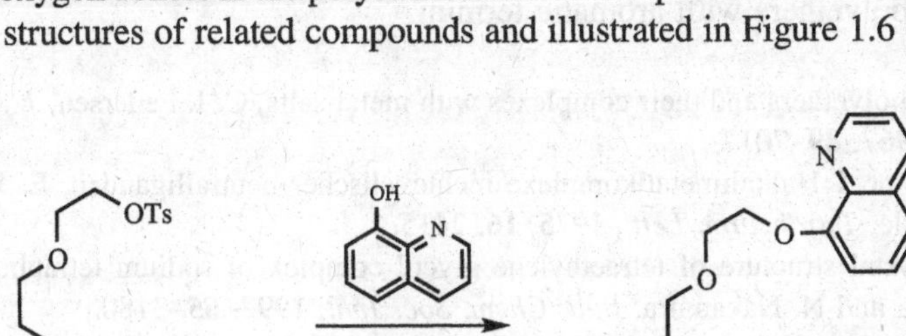

NaOH, H₂O

Figure 1.5 Syntheses of 1,9-bis(8-quinolinyloxy)-3,6-dioxanonane (**4**)

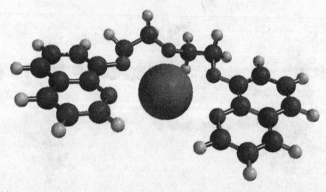

Figure 1.6 Simulation of a complex formed between a bis(quinoline) podand and sodium

Tosyl derivatives are widely used in polyether chemistry as the tosylate anion makes a good leaving group. Diiodopolyethers are often encountered too; however, iodide is no better as a leaving group under the conditions used in the examples given here. Tosylate derivatives have three clear advantages. First, they are often crystalline (though unfortunately not in the case of tetraethylene glycol ditosylate) and can therefore be obtained in high purity. Second, they add significantly to the mass of the parent compound (a mole of triethylene glycol weighs 150 g, its ditosylate derivative, 458 g) making transfer more accurate in the laboratory. Finally, the tosylate salts form as insoluble precipitates upon reaction with the alkali metal salts of alcohols and phenols in non-aqueous solvents. As a result, simple filtration followed by solvent removal is often all that is necessary to isolate the functionalized polyether.

Syntheses of ditosylates **2** and **3** are representative of the procedures required to make crystalline and non-crystalline polyether derivatives: continuation to bis(quinoline) **4** is typical of the method used to prepare a range of symmetrically substituted polyethers with aromatic termini.

[1] Cyclic polyethers and their complexes with metal salts, C. J. Pedersen, *J. Am. Chem. Soc.*, 1967, **89**, 7017.

[2] Kristalline 1:1-alkalimetallkomplexe nichtcyclischer neutralliganden, E. Weber and F. Vögtle, *Tetrahedron Lett.*, 1975, **16**, 2415.

[3] The crystal structure of tetraethylene glycol complex of sodium tetraphenylborate, T. Ueda and N. Nakamura, *Bull. Chem. Soc. Jpn.*, 1992, **65**, 3180.

[4] Macrocycle complexation chemistry 35. Survey of the complexation of the open chain 15-crown-5 analogue tetraethylene glycol with the lanthanide chlorides, R. D. Rogers, R. D. Etzenhouser, J. S. Murdoch and E. Reyes, *Inorg. Chem.*, 1991, **30**, 1445.

[5] Helicate self-organization–positive cooperativity in the self assembly of double-helical metal-complexes, A. Pfeil and J.-M. Lehn, *J. Chem. Soc., Chem. Commun.*, 1992, 838.

[6] Synthesis of 'crown ether' bislactones using caesium carboxylates of pyridine and of benzene dicarboxylic acids, O. Piepers and R. M. Kellog, *J. Chem. Soc., Chem. Commun.*, 1978, 383.

[7] Double stranded interwound infinite linear silver coordination network, B. Schmaltz, A. Jouaiti, M. W. Hosseini and A. De Cian, *Chem. Commun.*, 2001, 1242.

[8] Design and structural analysis of metallamacrocycles based on a combination of ethylene glycol bearing pyridine units with zinc, cobalt and mercury, P. Grosshans, A. Jouaiti, V. Bulach, J.-M. Planeix, M. W. Hosseini and N. Kyritsakas, *Eur. J. Inorg. Chem.*, 2004, 453.

[9] Pseudo cyclic ionophores: 'Binary-effect' of quinolinyloxy groups at both ends of oligoethylene glycols on the conformational stabilization of their complexes with alkali metal salts, R. Wakita, M. Miyakoshi, Y. Nakatsuki and M. Okahara, *J. Inclusion Phenom. Mol. Rec. Chem.*, 1991, **10**, 127.

[10] Open-chain polyethers. Influence of aromatic donor end groups on thermodynamics and kinetics of alkali metal ion complex formation, B. Tümmler, G. Maass, F. Vögtle, H. Sieger, U. Heimann and E. Weber, *J. Am. Chem. Soc.*, 1979, **101**, 2588.

[11] Macrocyclic oligo-ethers related to ethylene oxide, J. Dale and P. O. Kristiansen, *Acta Chem. Scand.*, 1972, **26**, 1471.

[12] Convenient and efficient tosylation of oligoethylene glycols and the related alcohols in tetrahydrofuran water in the presence of sodium hydroxide, M. Ouchi, Y. Inoue, Y. Liu, S. Nagamune, S. Nakamura, K. Wada and S. Hakushi, *Bull. Chem. Soc. Jpn.*, 1990, **63**, 1260.

[13] Structures of polyether complexes V. Molecular structure of bis(8-quinolyloxyethyl) ether-rubidium iodide, a linear polyether circularly embracing a metal ion, W. Saenger and B. S. Reddy, *Acta Cryst.*, 1979, **B35**, 56.

[14] Consequences of the Pedersen papers on crown type chemistry at Wurzburg and Bonn-Universities–from heteroaromatic crowns and podands to large molecular and crystalline cavities including multisite receptors, cascade molecules, chromoionophores, siderophores, surfactant-type, and extreme ligands, F. Vögtle and E. Weber, *J. Inclusion Phenom. Mol. Rec. Chem.*, 1992, **12**, 75.

Preparation of an isonicotinyl podand

1,19-Bis(isonicotinyloxy)-4,7,10,13,16-pentaoxaheptadecane (1)

Reagents	Equipment
Hexaethylene glycol	2-Necked round-bottomed flask (250 mL)
Triethylamine [FLAMMABLE]	Pressure equalized dropping funnel
Isonicotinyl chloride hydrochloride	Reflux condenser
Tetrahydrofuran (THF) [FLAMMABLE]	Heating/stirring mantle and stirrer bar
Diethyl ether [FLAMMABLE]	Inert atmosphere line
Distilled water	Glassware for filtration and work up
Hexane [FLAMMABLE]	
Dichloromethane [TOXIC]	
Magnesium sulphate	

Note: Wherever possible all steps of this synthesis should be carried out in a fume hood.

In a 250 mL two-necked round-bottomed flask, stir finely powdered isonicotinyl chloride hydrochloride (3.56 g, 20 mmol) in dry THF* (50 mL) under an inert atmosphere and add triethylamine (7.0 mL, 5.0 g, 50 mmol). Immediately the cream hydrochloride salt dissolves and a white precipitate of triethylamine hydrochloride forms. After 30 min add hexaethylene glycol (2.5 mL, 2.82 g, 10 mmol) in dry THF (30 mL) dropwise from a pressure-equalized dropping funnel. Once all the hexaethylene glycol solution has been added stir for a further 30 min then heat to reflux for 1 h. Allow the mixture to cool to room temperature and stir for a further 48 h. Filter the mixture and remove the THF under reduced pressure. (Place the precipitate in a fume hood as it may appear to smoke due to residual HCl. It should be dissolved in distilled water and neutralized with base prior to disposal.) Add distilled water (50 mL) to the residue and wash with hexane to remove any unreacted hexaethylene glycol. If the THF was dried with sodium hydride in mineral oil this process will also remove the oil. Extract the aqueous solution with dichloromethane (50 mL then twice with 25 mL). Dry the organic phase over anhydrous magnesium sulphate (*ca.* 1 g), filter and remove the solvent under reduced pressure to give the product, 1,19-bis(isonicotinyloxy)-4,7,10,13, 16-pentaoxaheptadecane (**1**), as an orange oil.

Yield: 3.8 g (80%); IR (v, cm^{-1}): 3440, 3035, 2875, 1955, 1730, 1410, 1285, 1120, 760, 710, 680; ^{1}H NMR (δ, ppm; CDCl$_3$) 8.8, 7.9 (dd, 8 H, ArH), 4.5 (t, 4 H, CH_2OC(O)Ar), 3.85 (t, 4 H, CH_2CH$_2$OC(O)Ar), 3.75–3.6 (m, 16 H, OCH_2CH$_2$O).

*For this reaction THF can be dried effectively using sodium hydride. Add about 1 g, in small portions, to a flask containing about 100 mL fresh reagent-grade THF, swirling after each addition, and repeat until no more effervescence is seen. Decant the solvent for use in the experiment but leave the remaining solid under enough solvent to stop it from drying out. To dispose of the residue (a mixture of sodium hydroxide and a small amount of unreacted sodium hydride) carefully add it a little at a time to a large volume of water. Once all the solid has dissolved check the pH of the solution and neutralize with dilute hydrochloric acid. Unless local safety regulations forbid it, the neutralized solution may be safely disposed of in a laboratory sink.

Preparation of polyethylene glycol ditosylates

Triethyleneglycol ditosylate (2)

Reagents
Triethylene glycol
Sodium hydroxide [CORROSIVE]
p-Toluenesulphonyl chloride

Equipment
2-Necked round-bottomed flask (2 L)
Pressure equalized addition funnel
Thermometer (−10 to 100 °C)

[CORROSIVE]
Tetrahydrofuran (THF) [FLAMMABLE]
Distilled water
Ethanol [FLAMMABLE]

Magnetic stirrer and stirrer bar
Ice bath
Glassware for recrystallization

Note: Work in a fume hood wherever possible, particularly when handling *p*-toluenesulphonyl chloride and when recrystallizing the product.

Prepare a solution of sodium hydroxide (40 g, 1 mol) in distilled water (200 mL) and cool to room temperature. Place the solution in a 2 L two-necked round-bottomed flask fitted with a thermometer and add a solution of triethylene glycol (56.5 g, 50 mL, 0.35 mol) in THF (200 mL) while stirring. Put the flask in an ice bath and cool to 0 °C. Place a solution of *p*-toluenesulphonyl chloride (145 g, 0.76 mol) in THF (200 mL) in a pressure-equalized addition funnel and add dropwise to the stirred glycol solution over 3 h or so. Carefully monitor the temperature of the solution and keep below 5 °C throughout the addition.* Once the addition of the *p*-toluenesulphonyl chloride solution is complete continue to stir the solution for a further 1 h below 5 °C. Pour onto a mixture of ice and water (250 g/250 mL) and continue to stir. After all the ice has melted filter the product, which forms as a white powder. Although the crude ditosylate could be used directly it is worthwhile recrystallizing from a minimum quantity of hot ethanol (*ca.* 2 mL per g). Note that the ditosylate will only dissolve when the ethanol is boiling, whereupon it becomes extremely soluble. Filter the boiling solution as quickly as possible to remove any insoluble matter, cool to room temperature, filter again to isolate the precipitate and dry thoroughly to remove residual ethanol. The product, triethylene glycol ditosyate (**2**), is obtained as a colourless microcrystalline solid or white powder.

Yield: 100+ g (60%); m.p.: 77–79 °C; IR (v, cm^{-1}): 3060, 2940, 1460, 1375, 1350, 1300, 1175, 1100; ^{1}H NMR (δ, ppm; CDCl$_3$) 8.8, 7.3 (dd, 8 H, Ar*H*), 4.2 (m, 4 H, C*H$_2$*OSO$_2$), 3.7 (m, 4 H, C*H$_2$*CH$_2$OSO$_2$), 3.9–3.8 (m, 8 H, OC*H$_2$*CH$_2$O), 2.4 (s, 6 H, ArC*H$_3$*).

*This takes some patience and it is advisable to keep a good supply of ice at hand.

Tetraethyleneglycol ditosylate (3)

Reagents
Tetraethylene glycol
Sodium hydroxide [CORROSIVE]
p-Toluenesulphonyl chloride
  [CORROSIVE]

Equipment
Two-necked round-bottomed flask (2 L)
Pressure equalized addition funnel
Thermometer (−10 to 100°C)
Magnetic stirrer and stirrer bar

Tetrahydrofuran (THF) [FLAMMABLE]   Ice bath
Distilled water   Glassware for recrystallization
Dichloromethane [TOXIC]   Rotary evaporator
Calcium sulphate

Note: Work in a fume hood wherever possible, particularly when handling *p*-toluenesulphonyl chloride.

Prepare a solution of sodium hydroxide (40 g, 1 mol) in distilled water (200 mL) and cool to room temperature. Place the solution in a 2 L two-necked round-bottomed flask fitted with a thermometer and add a solution of tetraethylene glycol (68 g, 60 mL, 0.35 mol) in THF (200 mL) while stirring. Put the flask in an ice bath and cool to 0 °C. Place a solution of *p*-toluenesulphonyl chloride (145 g, 0.76 mol) in THF (200 mL) in a pressure-equalized addition funnel and add dropwise to the stirred glycol solution over 3 h or so. Carefully monitor the temperature of the solution and keep below 5 °C throughout.* Once the addition of the *p*-toluene-sulphonyl chloride solution is complete, continue to stir the solution for a further 1 h at below 5 °C. Pour onto a mixture of ice and water (250 g/250 mL) and continue to stir. When all the ice has melted, remove most of the THF by rotary evaporation and extract the product into dichloromethane (3 × 100 mL). Dry the dichloromethane extract over calcium chloride, filter and remove the solvent by rotary evaporation. The product, tetraethylene glycol ditosylate (**3**), is obtained as a colourless oil.

Yield: ~160 g (95+%); IR (v, cm^{-1}): 3060, 2930, 1460, 1375, 1350, 1300, 1175, 1120; ^{1}H NMR (δ, ppm; CDCl$_3$) 8.8, 7.3 (dd, 8 H, ArH), 4.2 (m, 4 H, CH_2OSO$_2$), 3.7 (m, 4 H, CH_2CH$_2$OSO$_2$), 3.9–3.8 (m, 12 H, OCH_2CH$_2$O), 2.4 (s, 6 H, ArCH_3).

*See compound **2**.

Preparation of a quinoline podand

1,9-Bis(8-quinolinyloxy)-3,6-dioxanonane (4)

Reagents   **Equipment**
Triethylene glycol ditosylate (**2**)   2-Necked round-bottomed flask (250 mL)
8-Hydroxyquinoline   Heating/stirring mantle and stirrer bar
Sodium hydride (50% oil suspension)   Pressure equalized addition funnel
[CORROSIVE; REACTS VIOLENTLY   Inert atmosphere line
  WITH WATER]   Rotary evaporator
Tetrahydrofuran (THF) [FLAMMABLE]   Glassware for column chromatography

Distilled water
Dichloromethane [TOXIC]
Anhydrous sodium sulphate
Acetone [FLAMMABLE]
Silica (for chromatography)

Note: Work in a fume hood wherever possible and exercise due caution when handling sodium hydride.

Under an inert atmosphere, prepare a suspension of sodium hydride (0.96 g, 20 mmol [50 per cent in mineral oil]) in dry THF* (50 mL) in a 250 mL two-necked round-bottomed flask fitted with a reflux condenser and pressure equalized addition funnel. Stir for 30 min under an inert atmosphere. Slowly add a solution of 8-hydroxyquinoline (2.90 g, 20 mmol) in dry THF (50 mL) to this through the addition funnel. Make up a solution of triethylene glycol ditosylate, **2**, (4.60 g, 10 mmol) in dry THF (50 mL), ensuring that no solids remain (filter if necessary) or the addition funnel may become blocked. Once the effervescence subsides following the formation of the sodium 8-hydroxyquinolinate salt, add the ditosylate solution and reflux for 24 h. After 24 h, allow the solution to cool to room temperature, filter off the precipitated sodium tosylate and remove the THF by rotary evaporation. Dissolve the residue in dichloromethane (30 mL) and wash with distilled water (3 × 30 mL). Dry the organic phase over anhydrous sodium sulphate, filter and remove dichloromethane by rotary evaporation to give the crude product, 1,9-bis(8-quinolinyloxy)-3,6-dioxanonane (**4**) as a pale brown oil. Further purification may be afforded by column chromatography (silica, elute with acetone/dichloromethane).

Yield: 2.0 g (50%); IR (v, cm^{-1}): 3060, 2950, 1600, 1460, 1375, 1175, 1125; ^{1}H NMR (δ, ppm; CDCl$_3$) 8.9 (dd, 2 H, Ar*H*), 8.1 (dd, 2 H, Ar*H*), 7.4 (m, 6 H, Ar*H*), 7.1 (d, 2 H, Ar*H*), 4.4 (t, 4 H, OC*H*$_2$CH$_2$OAr), 4.1 (m, 4 H, OCH$_2$C*H*$_2$OAr), 3.8 (s, 4 H, OC*H*$_2$C*H*$_2$O).

*See compound **1** for drying method.

1.2 Rigid Components from Schiff Bases

While the flexibility of polyethers may be advantageous in many instances, it is often necessary to have a greater degree of preorganization in the ligands in order to entice guests, usually transition metals, into complex formation. Examples of this type include sexipyridine ligands used by Constable for metal complexation within a double helix [1] and Sauvage's copper-assisted formation of molecular trefoil knots from ligands derived from 1,10-phenanthroline [2].

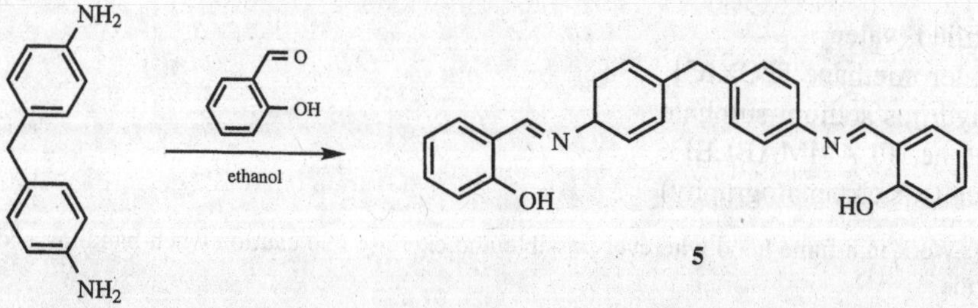

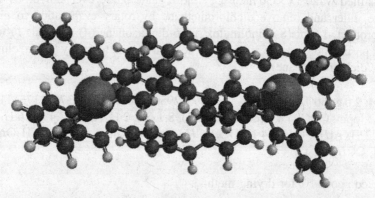

Figure 1.7 Synthesis of *N,N'*-(4,4'-methylenebiphenyl)bis(salicylideneimine) (**5**)

Hannon's group has prepared a range of rigid Schiff base ligands based on inexpensive starting materials [3]. The synthesis of these compounds is based on the addition of an aldehyde to a diamine (a method long used by coordination chemists to prepare ligands such as salens) with convergent binding sites, as shown in Figure 1.7. Coordination to transition metals results in highly coloured triple helical complexes containing two metal centres as can be seen in a simulation of Hannon's Ni_2L_3 complex, where L = bis(4-(2-pyridylmethyliminephenyl)methane) (Figure 1.8). Here, the diamine has divergent functionality leading to an extended ligand with an enforced twist.

Figure 1.8 Simulation of Hannon's triple helical Ni_2L_3 complex

A single example is given here. Originating from Hannon's work on Schiff base compounds, but substituting salicylaldehyde for pyridine-2-carbaldehyde, compound **5** is extremely easy to prepare and can be used to bind various transition metals. Multifunctional ligands, with the potential for extended hydrogen bonding or metal mediated polymerization may be prepared from 1, 3- or 1, 4-dihydroxybenzaldehyde. The general method given will work for most combinations of methylenedianiline and aromatic aldehydes.

[1] Oligopyridines as helicating ligands, E. C. Constable, *Tetrahedron*, 1992, **48**, 10013.
[2] A synthetic molecular trefoil knot, C. O. Dietrich-Buechecker and J.-P. Sauvage, *Angew. Chem. Int. Ed. Engl.*, 1989, **28**, 189.
[3] An inexpensive approach to supramolecular architecture, M. J. Hannon, C. L. Painting, A. Jackson, J. Hamblin and W. Errington, *Chem. Commun.*, 1997, 1807.

Preparation of a Schiff base helicate component

N,N'-(4,4'-Methylenebiphenyl)bis(salicylideneimine) (5)

Reagents
4,4-Methylenedianiline
Salicylaldehyde
Ethanol [FLAMMABLE]

Equipment
Conical flask (100 mL)
Magnetic stirrer and stirrer bar

Note: This compound can be prepared in a well-ventilated laboratory but will stain easily: take care when handling the product.

Place 4,4-methylenedianiline (1.6 g, 8 mmol) and ethanol (25 mL) in a 100 mL conical flask equipped with a stirrer bar. Stir at room temperature for about 20 min or until all the solid has dissolved. Add a solution of salicylaldehyde (1.9 g, 1.7 mL, 15 mmol) in ethanol (25 mL) and continue to stir. The solution turns an intense yellow colour before the product precipitates over 30 min as a yellow powder. The product is isolated in quantitative yield by filtration and washed with small quantities of ethanol to remove any unreacted starting materials. *N,N'*-(4,4'-methylenebiphenyl)bis(salicylideneimine) (**6**) is isolated as a yellow microcrystalline powder. Further purification should not be necessary, however, the product can be recrystallized from a minimum volume of boiling ethanol.

Yield: 3.2 g (quantitative); m.p.: 195–198 °C; IR (v, cm^{-1}): 3020, 1615, 1595, 980, 865; ^{1}H NMR (δ, ppm; CDCl$_3$/DMSO-d_6) 13.3 (s, 2 H, ArO*H*), 8.95 (s, 2 H, —NC*H*Ar), 7.55–7.2 (m, 10 H, Ar*H*), 6.85 (m, 4 H, Ar*H*), 3.25 (s, 2 H, ArC*H*$_2$Ar).

1.3 Flexible Tripods

The podands described in Section 1.1 are essentially linear molecules that encapsulate metals through conformational changes which allow electron donor atoms to converge upon the guest. Conformational change in solution requires a rearrangement of the solvation sphere around the ligand as well as the physical change in the ligand's conformation. While this is not necessarily an energy intensive process, ligands in which the donor groups are already preorganized for the guests will always be more effective complexants for metals, particularly where high binding constants are required.

The tripodal molecules prepared here are derivatives of tris(2-aminoethyl)-amine, more commonly known as *tren*, a readily available compound well known for its ligating abilities [1]. The primary amine termini of the ligand are ideally suited to reaction with aromatic aldehydes, forming Schiff base derivatives easily, and the central nitrogen may later act as a donor atom [2,3]. As the three amine termini diverge from the central nitrogen, trisubstitution occurs without one *tren* arm influencing the rate of reaction at any other arm. N^1,N^1,N^1-Tris(2-(2-aminoethylimino)methylphenol) has been used to bind a variety of metals, notably lanthanides, that can achieve high coordination numbers [4]. It has also been used to chelate ^{99m}Tc, providing yet another complex that can be considered for use in nuclear medicine [5]. Methoxy derivatives of this ligand can be used to bind two lanthanides at the same time, as the simulation in Figure 1.9 illustrates, to yield

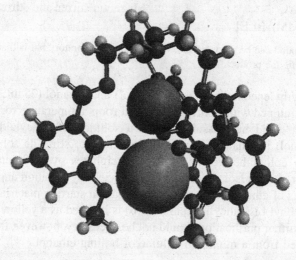

Figure 1.9 A bimetallic complex formed between a *tren*-derived ligand, gadolinium and neodymium

compounds with unusual magnetic properties [6]. Compounds based on the reduced forms of *tren*-derived ligands have also found widespread use as lanthanide chelators and as anion binding agents. The latter are discussed in the next section.

The syntheses of two very similar *tren*-derived podands are given here, N^1,N^1,N^1-tris(2-(2-aminoethylimino)methylphenol) (**6**) and N^1,N^1,N^1-tris(3-(2-aminoethylimino)methylphenol) (**7**), as shown in Figure 1.10. Both have phenolic termini: in the first they are in the 2-position and have the potential to converge on a guest, particularly a small transition metal, and in the second they are in the more divergent 3-position. By analogy with the extensive list of Schiff base ligands derived from ethylenediamine and aromatic aldehydes, many other tripodal ligands can be prepared using variations on the simple protocols given here.

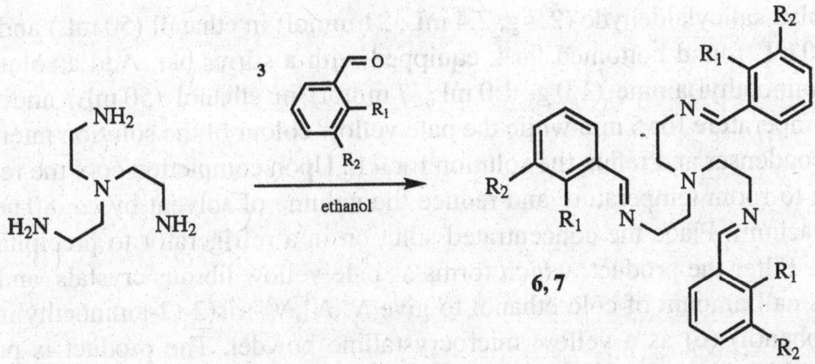

Figure 1.10 Syntheses of N^1,N^1,N^1-tris(2-(2-aminoethylimino)methylphenol) (**6**, $R_1 = OH$, $R_2 = H$) and N^1,N^1,N^1-tris(3-(2-aminoethylimino)methylphenol) (**7**, $R_1 = H$, $R_2 = OH$)

[1] Five-coordinated high-spin complexes of bivalent cobalt, nickel and copper with tris(2-dimethylaminoethyl)amine, M. Ciampolini and N. Nardi, *Inorg. Chem.*, 1966, **5**, 41.

[2] Low co-ordination numbers in lanthanide and actinide compounds. Part I. The preparation and characterization of tris{bis(trimethylsilyl)-amido}lanthanides, D. C. Bradley, J. S. Ghotra and F. A. Hart, *J. Chem. Soc., Dalton Trans.*, 1973, 1021.

[3] Potentially heptadentate ligands derived from tris(2-aminoethyl)amine (tren), A. Malek, G. C. Dey, A. Nasreen, T. A. Chowdhury and E. C. Alyea, *Synth. React. Inorg. Met.-Org. Chem.*, 1979, **9**, 145.

[4] Mononuclear lanthanide complexes of tripodal ligands: synthesis and spectroscopic studies, J.-P. Costes, A. Dupuis, G. Commenges, S. Lagrave and J.-P. Laurent, *Inorg. Chim. Acta*, 1999, **285**, 49.

[5] Synthesis of heptadentate (N_4O_3) amine–phenol ligands and radiochemical studies with technetium-99m, M. R. A. Pillai, K. Kothari, B. Mathew, N. K. Pilkwal, S. Jurisson, *Nucl. Med. Biol.*, 1999, **26**, 233.

[6] Unequivocal synthetic pathway to heterodinuclear (4f,4f') complexes: magnetic study of relevant (Ln^{III}, Gd^{III}) and (Gd^{III}, Ln^{III}) complexes, J.-P. Costes and F. Nicodème, *Chem. Eur. J.*, 2002, **8**, 3442.

Preparation of *tren*-derived tripods

N^1,N^1,N^1-*Tris(2-(2-aminoethylimino)methylphenol)* (**6**)

Reagents
Tris(2-aminoethyl)amine (*tren*)
Salicylaldehyde
Ethanol [FLAMMABLE]

Equipment
Round-bottomed flask (250 mL)
Heater/stirrer and stirrer bar
Reflux condenser
Rotary evaporator

Note: Wherever possible this reaction should be carried out in a fume hood.

Dissolve salicylaldehyde (2.7 g, 2.4 mL, 21 mmol) in ethanol (50 mL) and place in a 250 mL round-bottomed flask equipped with a stirrer bar. Add a solution of tris(2-aminoethyl)amine (1.0 g, 1.0 mL, 7 mmol) in ethanol (50 mL) and stir at room temperature for 5 min while the pale yellow colour of the solution intensifies. Add a condenser and reflux the solution for 4 h. Upon completion cool the reaction mixture to room temperature and reduce the volume of solvent by *ca.* 80 per cent under vacuum. Place the concentrated solution in a refrigerator to precipitate the product. Filter the product, which forms as pale yellow fibrous crystals, and wash with a small amount of cold ethanol to give N^1,N^1,N^1-tris(2-(2-aminoethylimino)-methylphenol) (**6**) as a yellow microcrystalline powder. The product is pure by NMR and should not require recrystallization.

Yield: 3.0 g (~95%); m.p.: 82–83 °C; IR (v, cm^{-1}): 3055, 2940, 2900, 2820, 1635, 1615, 1585, 1500, 1280, 755; ^{1}H NMR (δ, ppm; CDCl$_3$) 13.8 (s, 3 H, ArO*H*), 7.8 (s, 3 H, ArC*H*=N), 7.3 (m, 3 H, Ar*H*), 6.9 (d, 3 H, Ar*H*), 6.6 (m, 3 H, Ar*H*), 6.1 (m, 3 H, Ar*H*), 3.6 (t, 6 H, =NC*H*$_2$CH$_2$–), 2.8 (t, 6 H, =NCH$_2$C*H*$_2$–).

N^1,N^1,N^1-*Tris(3-(2-aminoethylimino)methylphenol)* (7)

Reagents

Tris(2-aminoethyl)amine (*tren*)
3-Hydroxybenzaldehyde
Methanol [FLAMMABLE]
Ice
Diethyl ether [FLAMMABLE]

Equipment

Round-bottomed flask (100 mL)
Magnetic stirrer and stirrer bar
Heat gun
Ice bath
Glassware for filtration

Note: Wherever possible this reaction should be carried out in a fume hood.

Prepare a solution of 3-hydroxybenzaldehyde (10.0 g, 82 mmol) in methanol (30 mL) in a 100 mL round-bottomed flask equipped with a stirrer bar. The mixture may require a little gentle warming with a heat gun to aid dissolution. Add a solution of tris(2-aminoethyl)amine (4.0 g, 7 mmol) in methanol (50 mL) and stir at room temperature for 10 min. The reaction is slightly exothermic; upon completion the reaction mixture cools to room temperature. Unusually for a Schiff base condensation of this type there is very little change in colour. Place an ice bath under the reaction vessel and stir until the product precipitates as a cream solid. Filter the product and wash with a small amount of cold ethanol to give N^1,N^1,N^1-tris(3-(2-aminoethylimino)methylphenol) (**7**) as a cream microcrystalline powder.

Yield: 3.4 g (quantitative); m.p.: 172–174 °C; IR (v, cm^{-1}): 2925, 2855, 2595, 1645, 1585, 1455, 1295; ^{1}H NMR (δ, ppm; CDCl$_3$) 9.5 (broad s, 3 H, ArO*H*), 8.1 (s, 3 H, ArC*H*=N), 7.2 (m, 3 H, Ar*H*), 7.1 (s, 3 H, Ar*H*), 7.0 (d, 3 H, Ar*H*), 6.8 (m, 3 H, Ar*H*), 3.6 (t, 6 H, =NC*H*$_2$CH$_2$–), 2.8 (t, 6 H, =NCH$_2$C*H*$_2$–).

1.4 Simple Anion Hosts

The natural world is concerned with binding anionic species just as much as cations and neutral molecules. It will not have escaped the notice of even the least biologically minded chemist that both DNA and RNA are polyanions. Vital biological processes involve the recognition and transport of anions such as sulphate, phosphate, carbonate and chloride. Impairment of the mechanisms by which these processes occur can have devastating effects for the organism concerned. For example, the deletion of three nucleotides from a particular gene results in the omission of a single phenylalanine from a transcribed protein. As a result the transmembrane transport of chloride across cells that form the lining of the lung is greatly reduced. This apparently trivial example is responsible for over 70 per cent of the incidents of cystic fibrosis, the most common lethal genetic defect in the Caucasian population. Another example of the importance of anion recognition is in the movement of phospholipids such as those with negatively charged phosphatidylserine head groups. Nature effects this with proteins that incorporate strongly bound calcium cations together with convergent recognition sites for carbonyl and amide groups [1].

A common motif for biological anion binding is the arginine residue which contains a terminal guanidinium group. This group is able to remain protonated over the entire physiological pH range and therefore provides an ideal mono-dentate or bidentate anion binding site. This motif is seen in the reversible phosphorylation of tyrosine. Protein tyrosine phosphatases that contain several convergent arginine residues are responsible for removing terminal phosphate groups from phosphotyrosine residues on proteins. The reverse process, transfer of a terminal phosphate from a nucleoside triphosphate to a tyrosine residue within a protein, is carried out by protein tyrosine kinases. The kinases were first identified in 1980 [2] and the phosphatases in 1988 [3]. Imbalance between the two processes has been implicated in many disease states including type-2 diabetes. One form of protein tyrosine phosphatase, YopH, is secreted by the bacterium *Yersinia pestis* responsible for the bubonic plague and may be a critical component in the virulence of that infection as it can translocate from the bacterium to the host organism [4]. Given the widespread biological occurrence of arginine as an anion-binding motif, it is no surprise that unnatural anion receptors have been designed to incorporate the related guanidinium cation, such as those in Figure 1.11. Guanidinium-based receptors have been prepared by several groups, most notably those of Lehn [5,6] and Schmidtchen [7]. Two excellent reviews are available covering general aspects of abiotic guanidinium groups in molecular recognition [8] and the more specific interactions between guanidinium groups and the oxyanions of phosphorus and sulphur [9].

The earliest anion-binding ligands to be prepared were those of Park and Simmonds [10,11] whose katapinands were similar to Lehn's later, and better known, cryptands but lacked coordinating oxygen atoms in the linkages connecting the

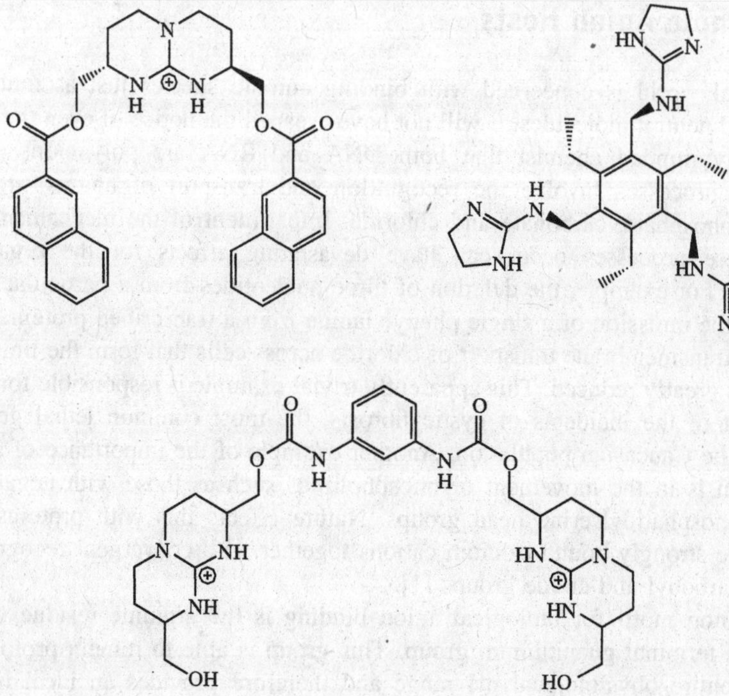

Figure 1.11   Guanidinium-based receptors

nitrogen atoms (Figure 1.12). In the diprotonated form this ligand was able to encapsulate chloride as shown by the X-ray structure of the complex. Graf and Lehn later reported that the tetraprotonated form of a tricyclic encapsulating ligand was able to bind both fluoride and chloride [12]. Similar compounds have been used to complex ATP through recognition of the polyphosphate residue by protonated regions in mixed oxa- and azacrown ethers that incorporate a pendent acridine group which is believed to π-stack with adenosine [13]. Many polyammonium-containing ligands, mostly macrocycles, have since been prepared. A particular design twist was incorporated by Schmidtchen who prepared molecular tetrahedra containing quaternary ammonium groups at the vertices [14].

An alternative approach is to design a proton-rich cavity that is flexible yet predisposed to bind to a particular anion. Examples of this type include the metal-containing polyazacryptands reported by Lehn [15] and Nelson [16] that encapsulate guests such as succinate and nitrate. A more flexible approach has been taken by Reinhoudt [17], Beer [18] and Bowman-James [19] who have used tris(aminoethyl)amine, *tren*, as the basis for preorganized anion binding and whose derivatives as shown in Figure 1.13. Most recently this theme has been extended to include *tren*-derived, Schiff-base podands that have subsequently been reduced to form trisamines capable of binding anions, from phosphate to bromide, with a range of specificities [20]. The preparation of reduced Schiff-base podands is not

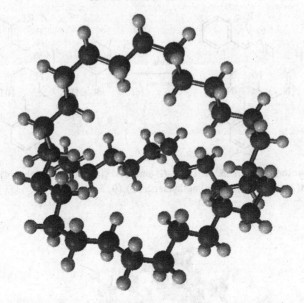

Figure 1.12 A katapinand in the 'in–in' conformation

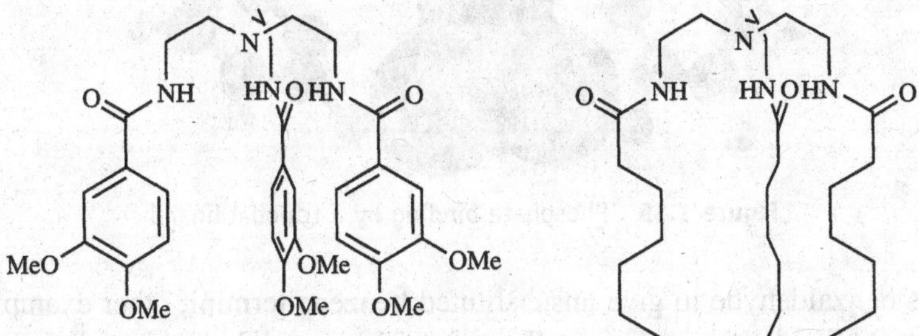

Figure 1.13 *Tren*-derived anion-binding podands

new [21–23], however, their application to the field of anion recognition is becoming more widespread. One example given here uses potassium borohydride to reduce the podand N^1,N^1,N^1-tris(2-(2-aminoethylimino)methylphenol) (**6**) and the other gives a one-pot synthesis of N^1,N^1,N^1-tris((2-aminoethylamino)methyl-benzene) (**9**). The synthesis of the former can be found in a preceding section. Note that the reduction of triphenol **7** yields a derivative that can only be extracted into chloroform (dichloromethane is not polar enough) and is hydroscopic. As a result redissolution of the isolated product is difficult. The reduced ligands, shown in Figure 1.14, have the ability to bind a variety of guests but, due to the presence of amine groups, are ideal for tetrahedral oxyanions such as phosphate (Figure 1.15). Many variations are possible with this class of podands, most obviously in the aromatic substitution pattern. The most basic example [20]

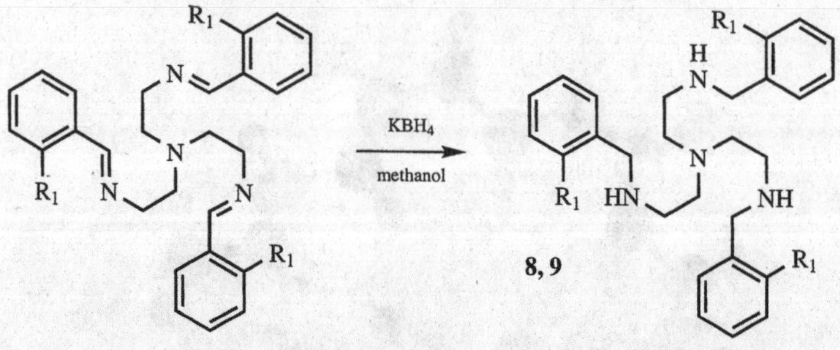

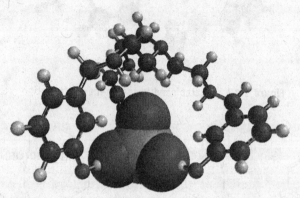

Figure 1.14 Syntheses of N^1,N^1,N^1-tris(2-(2-aminoethylamino)methylphenol) (**8**, $R_1 = OH$) and N^1,N^1,N^1-tris(2-aminoethylamino)methylbenzene) (**9**, $R_1 = H$)

Figure 1.15 Phosphate binding by a tripodal ligand

employs benzaldehyde to give unsubstituted benzene termini; other examples use hydroxynaphthaldehyde and *o*-vanillin to introduce naphthyl and methoxybenzene functionality, respectively. The amine core of these tripodal ligands also has the potential to aid in ligand preorganization by binding metal ions. The resulting cationic species will attract anions of the correct size and complementary geometry. The copper complex of a more complex reduced *tren*-derivative, incorporating benzylamine termini, has been shown to bind a variety of anionic guests [24]. In summary, the reduced forms of the *tren*-derived ligands can complex a remarkable range of species from simple anions to lanthanides. It would be surprising if they did not also complex a range of transition and main group metals.

[1] Crystal structure of a calcium-phospholipid binding domain from cytosolic phospholipase A2, O. Perisic, S. Fong, D. E. Lynch, M. Bycroft and R. L. Williams, *J. Biol. Chem.*, 1998, **273**, 1596.

[2] Transforming gene product of Rous sarcoma virus phosphorylates tyrosine, T. Hunter and B. M. Sefton, *Proc. Natl. Acad. Sci. U S A*, 1980, **77**, 1311.

[3] Purification of the major protein-tyrosine-phosphatases of human placenta, N. K. Tonks, C. D. Diltz and E. H. Fischer, *J. Biol. Chem.*, 1988, **263**, 6722.

[4] *Yersinia* effectors target mammalian signalling pathways, S. J. Juris, F. Shao and J. E. Dixon, *Cell. Microbiol.*, 2002, **4**, 201.

[5] Anion receptor molecules. Synthesis and some anion binding properties of macrocyclic guanidinium salts, B. Dietrich, T. M. Fyles, J.-M. Lehn, L. G. Pease and D. L. Fyles, *J. Chem. Soc., Chem. Commun.*, 1978, 934.

[6] Anion coordination chemistry: polyguanidinium salts as anion complexones. B. Dietrich, D. L. Fyles, T. M. Fyles and J.-M. Lehn, *Helv. Chim. Acta*, 1979, **62**, 2763.

[7] Host-guest bonding of oxoanions to guanidinium anchor groups, G. Müller, J. Riede and F. P. Schmidtchen, *Angew. Chem. Int. Ed. Engl.*, 1988, **100**, 751.

[8] Abiotic guanidinium containing receptors for anionic species, M. D. Best, S. L. Tobey and E. V. Anslyn, *Coord. Chem. Rev.*, 2003, **240**, 3.

[9] Noncovalent binding between guanidinium and anionic groups: focus on biological- and synthetic-based arginine/guanidinium interactions with phosph[on]ate and sulf [on]ate residues, K. A. Shug and W. Lindner, *Chem. Rev.*, 2005, **105**, 67.

[10] Macrobicyclic amines I. Out-in isomerism of (1,k+2)diazabicyclo[k.l.m]alkanes, H. E. Simmonds and C. H. Park, *J. Am. Chem. Soc.*, 1968, **90**, 2428.

[11] Macrobicyclic amines II. Out-out in-in prototropy in (1,k+2)diazabicyclo[k.l.m] alkane ammonium ions, C. H. Park and H. E. Simmonds, *J. Am. Chem. Soc.*, 1968, **90**, 2431.

[12] Anion cryptates: highly stable and selective macrotricyclic anion inclusion complexes, E. Graf and J.-M. Lehn, *J. Am. Chem. Soc.*, 1976, **98**, 6403.

[13] Supramolecular catalysis in the hydrolysis of ATP facilitated by macrocyclic poly-amines—mechanistic studies, M. W. Hosseini, J.-M. Lehn, L. Maggioria, K. B. Mertes and M. P. Mertes, *J. Am. Chem. Soc.*, 1987, **109**, 537.

[14] Inclusion of anions in macrotricyclic quaternary ammonium salts, F. P. Schmidtchen, *Angew. Chem. Int. Ed. Engl.*, 1977, **89**, 720.

[15] Polyaza macrobicyclic cryptands: synthesis, crystal structures of a cyclophane type macrobicyclic cryptand and of its dinuclear copper(I) cryptate, and anion binding features, J. Jazwinski, J.-M. Lehn, D. Lilienbaum, R. Ziessel, J. Guilhem and C. Pascard, *J. Chem. Soc., Chem. Commun.*, 1987, 1691.

[16] Chemistry in cages: dinucleating azacryptand hosts and their cation and anion cryptates, M. Arthurs, V. McKee, J. Nelson and R. M. Town, *J. Chem. Ed.*, 2001, **78**, 1269.

[17] Synthesis and complexation studies of neutral anion receptors, S. Valiyayeettil, J. F. J. Engbersen, W. Verboom and D. N. Reinhoudt, *Angew. Chem. Int. Ed. Engl.*, 1993, **32**, 900.

[18] Cooperative halide, perrhenate anion—sodium cation binding and pertechnetate extrac-tion and transport by a novel tripodal tris(amido benzo-15-crown-5) ligand, P. D. Beer, P. K. Hopkins and J. D. McKinney, *Chem. Commun.*, 1999, 1253.

[19] Novel structural determination of a bilayer network formed by a tripodal lipophilic amide in the presence of anions, A. Danby, L. Seib, N. W. Alcock, K. Bowman-James, *Chem. Commun.*, 2000, 973.

[20] Anion binding with a tripodal amine, M. A. Hossain, J. A. Liljegren, D. Powell and K. Bowman-James, *Inorg. Chem.*, 2004, **43**, 3751.

[21] Synthesis and characterization of lanthanide [Ln(L)]$_2$ complexes of N$_4$O$_3$ amine phenol ligands with phenolate oxygen bridges – evidence for very weak magnetic exchange between lanthanide ions, S. Liu, L. Gelmini, S. J. Rettig, R. C. Thompson and C. Orvig, *J. Am. Chem. Soc.*, 1992, **114**, 6081.

[22] Bulky ortho 3-methoxy groups on N$_4$O$_3$ amine phenol ligands producing six-coordinate bis(ligand)lanthanide complex cations [Ln(H$_3$L)$_2$]$^{3+}$ (Ln = Pr, Gd; H$_3$L = tris(((2-hydroxy-3-methoxybenzyl)amino)ethyl)amine), S. Liu, L.-W. Yang, S. J. Rettig and C. Orvig, *Inorg. Chem.*, 1993, **32**, 2773.

[23] Mononuclear lanthanide complexes of tripodal ligands: synthesis and spectroscopic studies, J.-P. Costes, A. Dupuis, G. Commenges, S. Lagrave and J.-P. Laurent, *Inorg. Chim. Acta*, 1999, **285**, 49.

[24] Studies into the thermodynamic origin of negative cooperativity in ion-pairing molecular recognition, S. L. Tobey and E. V. Anslyn, *J. Am. Chem. Soc.*, 2003, **125**, 10963.

Reduction of *tren*-derived tripods

N^1,N^1,N^1-*Tris(2-(2-aminoethyamino)methylphenol) (8)*

Reagents

N^1,N^1,N^1-Tris(2-(2-aminoethylimino)-
methylphenol) (**7**)
Potassium borohydride [CORROSIVE;
REACTS VIOLENTLY WITH WATER]
Methanol [FLAMABLE]
Ammonium chloride
(2 M, aqueous solution)
Chloroform [TOXIC; CARCINOGEN]
Distilled water
Anhydrous magnesium sulphate

Equipment

Round-bottomed flask (250 mL)
Magnetic stirrer and stirrer bar
Heat gun
Rotary evaporator
Glassware for extraction and work up

Note: This reaction should be carried out in a well-ventilated laboratory. Exercise due care when handling potassium borohydride.

Dissolve N^1,N^1,N^1-tris-(2-(2-aminoethylimino)methylphenol) (**7**) (1 g, 2.2 mmol) in methanol (100 mL) in a 250 mL round-bottomed flask equipped with a stirrer bar. Gentle heating may be necessary to effect complete dissolution. Add potassium borohydride (0.4 g, 7.4 mmol) in small portions to the bright yellow solution. During the addition the solution completely decolorizes. Continue to stir at room temperature for a further 1 h after the last portion of potassium borohydride has been added. Upon completion remove the solvent under vacuum to leave a white solid. Add a solution of ammonium chloride (20 mL of a 2 M aqueous solution) to the crude product, cautiously at first, to ensure all the potassium borohydride has been consumed. Extract with chloroform (50 mL then 2 × 25 mL), wash with distilled water (30 mL) and dry the organic phase with magnesium sulphate (*ca.* 1 g). Filter the solution and remove the solvent under vacuum to give N^1,N^1,N^1-tris(2-(2-aminoethylamino)methylphenol) (**8**) as a colourless, hydroscopic oil that solidifies when held under high vacuum for several hours.

Yield: 0.8 g (∼80%); m.p.: 45–46 °C; IR (v, cm^{-1}): 3310, 3010, 2825, 1630, 1590, 1490, 1255; ^{1}H NMR (δ, ppm; DMSO-d_6) 7.1–7.0 (m, 6 H, Ar*H*), 6.8–6.7 (m, 6 H, Ar*H*), 3.8 (s, 6 H, NC*H*$_2$Ar), 2.6–2.4 (m, 12 H, NC*H*$_2$C*H*$_2$N).

N^1,N^1,N^1-*Tris((2-aminoethylamino)methylbenzene)* (9)

Reagents	Equipment
Tris(2-aminoethyl)amine (*tren*)	Round-bottomed flask (250 mL)
Benzaldehyde	Magnetic stirrer and stirrer bar
Methanol [FLAMMABLE]	Calcium chloride guard tube
Potassium borohydride [CORROSIVE;	Glassware for extraction and work up
REACTS VIOLENTLY WITH	Rotary evaporator
WATER]	
Sodium hydroxide [CORROSIVE]	
Chloroform [TOXIC; CARCINOGEN]	
Distilled water	
Anhydrous magnesium sulphate	

Note: This reaction should be carried out in a well-ventilated laboratory. Exercise due care when handling potassium borohydride.

Prepare a solution of tris(2-aminoethyl)amine (1.95 g, 2.0 mL, 13 mmol) in methanol (75 mL) in a 250 mL round-bottomed flask equipped with a stirrer bar. Dissolve benzaldehyde (4.2 g, 4.0 mL, 40 mmol) in methanol (25 mL), add slowly to the *tren* solution, stopper the flask and stir at room temperature for 24 h while the pale yellow colour intensifies. Upon completion, remove the stirrer bar and reduce the volume of solvent by *ca.* 50 per cent under vacuum. To this methanolic solution of N^1,N^1,N^1-tris(2-(2-aminoethylimino)methylbenzene) add potassium borohydride (4.9 g, 90 mmol) in small portions while stirring vigorously. After the final addition, fit a calcium chloride guard tube. A slightly exothermic reaction ensues and the mixture effervesces for several hours. Stir at room temperature for 24 h while the solution fades in intensity and a pale yellow solid precipitates. Cautiously add an aqueous solution of sodium hydroxide (8 g in 50 mL) and stir for a further 1 h. Extract the resulting white emulsion with chloroform (1 × 100 mL then 2 × 50 mL), wash with distilled water (50 mL), dry with magnesium sulphate, filter and remove solvent using a rotary evaporator. Note that the addition of water leads to the formation of a white suspension; addition of magnesium sulphate clarifies the solution. The product, N^1,N^1,N^1-tris((2-aminoethylimino)methylben-zene) (**9**) is isolated as a yellow oil and is suitably pure for further experimentation.

Yield: 4.9 g (90%); IR (v, cm^{-1}): 3025, 2815, 1450, 1050, 730, 695; ^{1}H NMR (δ, ppm; CDCl$_3$) 7.4–7.2 (m, 15 H, ArH), 3.7 (s, 6 H, ArCH_2), 2.7 (t, 6 H, NHCH_2CH$_2$N), 2.5 (t, 6 H, NHCH$_2$CH_2N).

1.5 Rigid Platforms

Other 'molecular tripods' have been prepared, such as those based on 1,3,5-trisubstituted cyclohexane, including a wide range of ligands that have evolved from Kemp's triacid, in particular those devised by Rebek [1] and the related work of the Walton group [2,3]. This motif is of particular interest to those investigating self-replication of small molecules. N,N',N''-Trisubstituted triazacyclononane, [9]aneN$_3$, also makes an excellent platform for further functionalization. The binding pockets present in these compounds make their metal complexes interesting mimics for a range of enzymes, particularly those incorporating zinc or copper. Some representative examples of rigid tripodal ligands are shown in Figure 1.16.

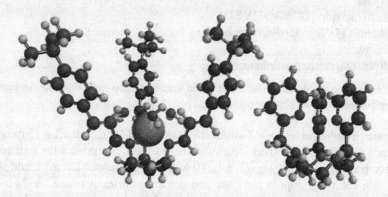

Figure 1.16 Rigid tripodal ligands based on 1,3,5-trisubstituted cyclohexane (left) and 1,3,9-triazacyclononane (right)

In addition to the tripodal structural motifs mentioned above, one obvious choice to imbue a ligand with threefold symmetry is to use 1,3,5-trisubstituted benzene as a base. The formation of the alternating conformation for substituents on a hexasubstituted benzene is a well-known phenomenon: an early example where this was used in a supramolecular context was by Harshorn and Steel in 1996 [4]. Subsequently the principle has been used widely particularly in the detection of anions [5–8]. The parent 1,3,5-tris(bromomethyl)-2,4,6-triethylben-zene is fairly complex to prepare [9], however, the corresponding mesitylene derivative, 1,3,5-tris(bromomethyl)-2,4,6-trimethylbenzene, can be made on a multigram scale in yields approaching 100 per cent [10]. From this derivative Sato [11], and later Howarth [12], were able to prepare tripodal anion receptors. An excellent review was published by Anslyn, which charts the development of these compounds up to 2002 [13]. More recently, Itoh has shown that a variety of different topologies may result from the reaction of a related 1,3,5-tris[2-(pyridin-2-yl)ethyl]-2,4,6-triethylbenzene with copper, zinc or palladium [14].

The example given here is derived from 1,3,5-tri(bromomethyl)-2,4,6-trimethyl-benzene (10), as it is significantly easier to prepare than the 2,4,6-triethylbenzene analogue (Figure 1.17). The choice of benzene as a base for the tripod gives a

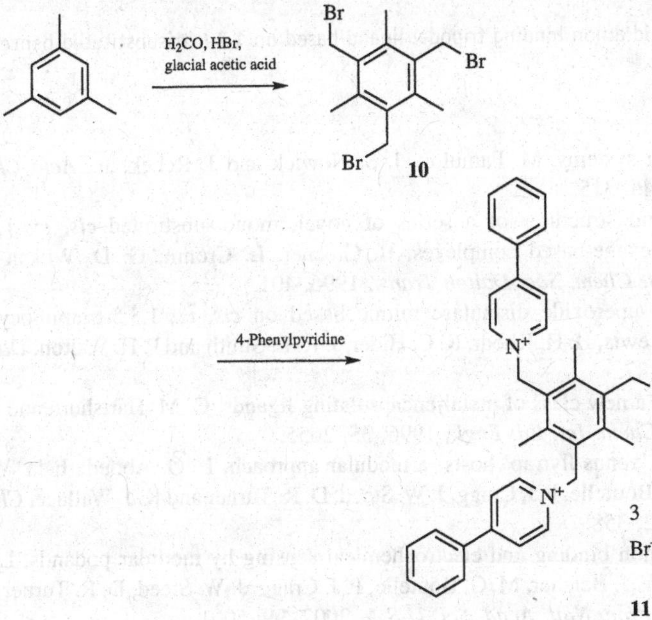

Figure 1.17 Synthesis of tripodal ligand **11** via tribromide **10**

sterically hindered product in which the ideal structure has 1,3,5-substituents and 2,4,6-substituents directed in opposite directions from face of the aromatic ring. Coordinating groups can then be focused in a convergent fashion towards a point perpendicular to the centre of the aromatic ring to give a well-defined binding site as shown in Figure 1.18. The range of tripyridyl compounds that could be prepared from 1,3,5-tri(bromomethyl)trimethylbenzene is probably only limited by the solubilities of the pyridine derivatives in dichloromethane. The method described for the tri(4-phenylpyridine) derivative (**11**) should be applicable to numerous variations on the example given.

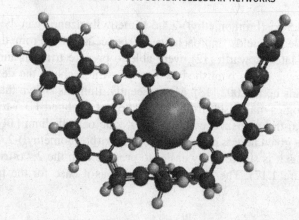

Figure 1.18 A rigid anion binding tripodal ligand based on 1,3,5-trisubstituted benzene

[1] Self-replicating systems, M. Famulok, J. S. Nowick and J. Rebek, Jr., *Acta Chem. Scand.*, 1992, **46**, 315.

[2] Preparations and structures of a series of novel, mono-substituted *cis*, *cis*-1,3,5-triaminocyclohexane-based complexes, B. Greener, L. Cronin, G. D. Wilson and P. H. Walton, *J. Chem. Soc. Dalton Trans.*, 1996, 401.

[3] A manganese superoxide dismutase mimic based on *cis*, *cis*-1,3,5-triaminocyclohexane, E. A. Lewis, H. H. Khodr, R. C. Hider, J. R. L. Smith and P. H. Walton, *Dalton Trans.*, 2004, 187.

[4] Coelenterands: a new class of metal-encapsulating ligands, C. M. Hartshorn and P. J. Steel, *Angew. Chem. Int. Ed. Engl.*, 1996, **35**, 2655.

[5] Anion sensing 'venus flytrap' hosts: a modular approach, L. O. Abouderbala, W. J. Belcher, M. G. Boutelle, P. J. Cragg, J. W. Steed, D. R. Turner and K. J. Wallace, *Chem. Commun.*, 2002, 358.

[6] Cooperative anion binding and electrochemical sensing by modular podands, L. O. Abouderbala, W. J. Belcher, M. G. Boutelle, P. J. Cragg, J. W. Steed, D. R. Turner and K. J. Wallace, *Proc. Natl. Acad. Sci. U S A*, 2002, **99**, 5001.

[7] A synthetic receptor selective for citrate, A. Metzger, V. M. Lynch and E. V. Anslyn *Angew. Chem. Int. Ed. Engl.*, 1997, **36**, 862.

[8] A molecular flytrap for the selective binding of citrate and other tricarboxylates in water, C. Schmuck and M. Schwegmann, *J. Am. Chem. Soc.*, 2005, **127**, 3373.

[9] Synthesis of 1,3,5-tris(bromomethyl)-2,4,6-triethylbenzene – a versatile precursor to predisposed ligands, C. Walsdorff, W. Saak, S. Pohl, *J. Chem. Res. (M)*, 1996, 1601.

[10] A convenient procedure for bromomethylation of aromatic compounds. Selective mono-, bis-, or trisbromomethylation, A. W. van der Made and R. H. van der Made, *J. Org. Chem.*, 1993, **58**, 1262.

[11] A new tripodal anion receptor with C-H···X⁻ hydrogen bonding, K. Sato, S. Arai and T. Yamagishi, *Tetrahedron Lett.*, 1999, **40**, 5219.

[12] A homochiral tripodal receptor with selectivity for sodium (*R*)-2-aminopropionate over sodium (*S*)-2-aminopropionate, J. Howarth and N. A. Al-Hashimy, *Tetrahedron Lett.*, 2001, **42**, 5777.

[13] 1,2,3,4,5,6-Functionalised, facially segregated benzenes–exploitation of sterically predisposed systems in supramolecular chemistry, G. Hennrich and E. V. Anslyn, *Chem. Eur. J.*, 2002, **8**, 2219.

[14] Supramolecular and coordination polymer complexes supported by a tripodal tripyridine ligand containing a 1,3,5-triethylbenzene spacer, H. Ohi, Y. Tachi and S. Itoh, *Inorg. Chem.*, 2004, **43**, 4561.

Preparation of 1,3,5-trisubstituted benzene tripods

1,3,5-Tri(bromomethyl)-2,4,6-trimethylbenzene (10)

Reagents

1,3,5-Trimethylbenzene (mesitylene)

Paraformaldehyde [TOXIC; CARCINOGENIC]

Glacial acetic acid [CORROSIVE]

Hydrogen bromide in acetic acid (31%) [CORROSVE]

Distilled water

Diethyl ether [FLAMMABLE]

Equipment

Round-bottomed flask (100 mL)

Heating mantle/stirrer and stirrer bar

Reflux condenser

Glassware for filtration

Note: This reaction should be carried out in a fume hood.

Add mesitylene (6.0 g, 50 mmol), paraformaldehyde (5 g, 170 mmol) and glacial acetic acid (25 mL) to a 100 mL round-bottomed flask and stir at room temperature. Add a solution of hydrogen bromide in acetic acid (35 mL, 31 per cent v/v), heat to 90 °C and stir for 12 h. At this temperature the paraformaldehyde dissolves to give an orange solution. When the reaction is complete a white solid remains. Pour this into water (100 mL) and rinse out the reaction flask with more water (100 mL). Stir vigorously to break up the precipitate and filter to give 1,3,5-tri(bromomethyl)trimethylbenzene (**10**) as a white solid. Further purification can be afforded by stirring the solid in boiling diethyl ether (200 mL) and filtering. Although this process reduces the yield slightly it also removes coloured impurities. Note that this compound starts to decompose if left under ambient conditions for more than a week and should be used immediately to prepare derivative **11**.

Yield: 14 g (70%); m.p.: 186 °C; IR (v, cm^{-1}): 3025, 1560, 1080, 1010, 785, 575; ^{1}H NMR (δ, ppm; CDCl$_3$) 4.6 (s, 6 H, ArCH_2Br), 2.5 (s, 9 H, ArCH_3).

1,3,5-Tris[(4-phenylpyridine)methyl]-2,4,6-trimethylbenzene tribromide (11)

Reagents
1,3,5-Tri(bromomethyl)-
   trimethylbenzene (**10**)
4-Phenylpyridine
Dichloromethane [TOXIC]

Equipment
Round-bottomed flask (250 mL)
Pressure-equalized dropping funnel
Inert atmosphere line
Glassware for work-up

Note: This reaction should be carried out in a fume hood.

Stir a solution of 4-phenylpyridine (1.16 g, 7.5 mmol) in dichloromethane (50 mL) in a 250 mL round-bottomed flask at room temperature under an inert atmosphere. Add a solution of 1,3,5-tri(bromomethyl)trimethylbenzene (1 g, 2.5 mmol) in dichloromethane (50 mL) dropwise from a pressure-equalized dropping funnel. A white precipitate forms after an hour or so, however, it is worth stirring for a further 12 h to ensure the reaction goes to completion. When the reaction is complete, filter and wash with dichloromethane (25 mL) to give 1,3,5-tris[(4-phenylpyridine)methyl]-2,4,6-trimethylbenzene tribromide (**11**) as a white powder.

Yield: 2.1 g (95%); m.p.: >250 °C; IR (v, cm^{-1}): 3100, 3025, 1635, 1555, 1135, 850, 765, 715, 695; ^{1}H NMR (δ, ppm; CD$_3$OD) 8.85 (m, 6 H, Ar*H*), 8.45 (m, 6 H, Ar*H*), 8.05 (m, 6 H, Ar*H*), 7.65 (m, 9 H, Ar*H*), 4.9 (s, 6 H, C*H*$_2$), 2.6 (s, 9 H, C*H*$_3$).

2
Cyclic Synthons

2.1 Planar Macrocycles from Nature

The porphyrin motif, or a variation thereof, is found throughout nature where it is used to bind a variety of metal cations including magnesium (chlorophyll), manganese (photosystem II), cobalt (vitamin B_{12}), nickel (coenzyme F-430) and iron (haemoglobin and myoglobin) [1]. The structures of these bioinorganic systems have been a source of inspiration for supramolecular chemists for many years: it is instructive to note that some therapeutic compounds, such as Sessler's texaphyrins, have their origins in porphyrin mimicry [2]. From a mechanistic standpoint it is often necessary for bioinorganic chemists to synthesize analogues of the metal centred sites that exist naturally within enzymes or metalloproteins in order to elucidate the subtleties of metal-centred catalysis or substrate binding. To this end many porphyrin derivatives have been prepared for study and some have since been incorporated in supramolecular cages or transformed into biomimetic systems [3,4]. Tetraphenylporphyrin (**12**) was one of the earliest synthetic porphyrins to be reported and its synthesis remains the easiest to reproduce [5]. Several variations have been reported, however, the method published in 1967 is adequate for any researcher requiring a representative compound. The overall yield may appear to be low but it must be remembered that the product is formed by the condensation of four molecules of pyrrole and four of benzaldehyde without the need for a template, as shown in Figure 2.1. In addition to nuclear magnetic resonance (NMR) and infrared spectral analysis this compound is characterized by an intense absorbance at 415 nm in the visible region: the so-called Soret band. Preparation of metal complexes from the free ligand is also simple to achieve and many methods are available in the literature [6].

[1] *Bioinorganic Chemistry: Inorganic Elements in the Chemistry of Life – An Introduction and Guide*, W. Kaim and B. Schwederski, J. Wiley & Sons, Chichester, 1994.

A Practical Guide to Supramolecular Chemistry Peter J. Cragg
© 2005 John Wiley & Sons, Ltd

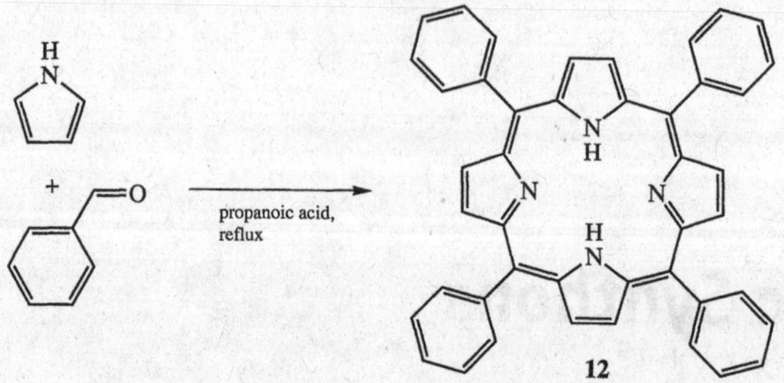

Figure 2.1 Synthesis of tetraphenylporphyrin (**12**)

[2] An 'expanded porphyrin': the synthesis and structure of a new aromatic pentadentate ligand, J. L. Sessler, T. Murai, V. Lynch and M. Cyr, *J. Am. Chem. Soc.*, 1988, **110**, 5586.

[3] Metal-driven self assembly of C_3 symmetry molecular cages, F. Felluga, P. Tecilla, L. Hillier, C. A. Hunter, G. Licini and P. Scrimin, *Chem. Commun.*, 2000, 1087.

[4] Self-assembly of pentameric porphyrin light-harvesting antennae complexes, R. A. Haycock, A. Yartsev, U. Michelsen, V. Sundström and C. A. Hunter, *Angew. Chem. Int. Ed. Engl.*, 2000, **39**, 3616.

[5] A simplified synthesis for meso-tetraphenylporphine, A. D. Adler, F. R. Longo, J. D. Finarelli, J. Goldmacher, J. Assour and L. Korsakoff, *J. Org. Chem.*, 1967, **32**, 476.

[6] Microscale synthesis and electronic absorption spectroscopy of tetraphenylporphyrin H_2(TPP) and metalloporphyrins Zn^{II}(TPP) and Ni^{II}(TPP), D. F. Marsh and L. M. Mink, *J. Chem. Ed.*, 1996, **73**, 1188.

Preparation of tetraphenylporphyrin

Tetraphenylporphyrin (12)

Reagents
Pyrrole [HARMFUL]
Propanoic acid [CORROSIVE]
Benzaldehyde [TOXIC]
Methanol [FLAMMABLE]

Equipment
Small-scale distillation apparatus
Heating/stirring mantles (25 and 250 mL) and stirring bars
Thermometer
Round-bottomed flasks (25 and 250 mL)
Reflux condenser
Stirrer bar
Pasteur pipettes
Büchner funnel and flask

Note: The distillation and porphyrin synthesis must be carried out in a fume hood as several of the reagents have a pungent smell.

Distil *ca.* 3 mL pyrrole from a 25 mL round-bottomed flask, collecting the colourless fraction that boils between 128 and 131 °C. If required this can be stored in a freezer for several days before use.

Carefully pour propanoic acid (75 mL) into a 250 mL round-bottomed flask containing a magnetic stirrer and fit a reflux condenser. Bring the acid to reflux (b.p. 141 °C) then simultaneously add the freshly distilled pyrrole (1.4 mL, 1.35 g, 20 mmol) and benzaldehyde (2.0 mL, 2.1 g, 20 mmol) down the condenser using two Pasteur pipettes. Continue to heat the mixture, which will rapidly change from colourless to dark brown, for a further 30 min. After 30 min allow the mixture to cool to room temperature and collect the deep purple product by suction filtration using a small Büchner funnel. Note that the acidic solution contains highly coloured pyrrole side products that will stain. It is strongly advised that gloves are worn during filtration. Wash thoroughly with methanol until the washings are colourless. Dry by suction and isolate the crystalline purple tetraphenylporphyrin (**12**) (Figure 2.2).

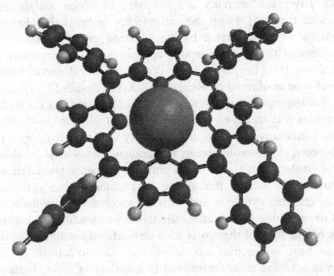

Figure 2.2 Simulated structure of the copper tetraphenylporphyrin complex

Yield: 0.30 g, (10%); m.p.: >250 °C; IR (v, cm^{-1}): 3125, 1595, 1560, 1350, 1175, 1000, 965, 805, 740; ^{1}H (δ, ppm; CDCl$_3$) 8.85 (s, 8 H, ArH), 8.2 (dd, 8 H, CH), 7.95 (m, 12 H, ArH), −2.75 (s, 2 H, NH); UV-vis (CHCl$_3$) 415 nm.

2.2 Artificial Planar Macrocycles – Phthalocyanines and Other Cyclic Systems

As is the case with so many chemical discoveries, phthalocyanines were first obtained through the accidental addition of a catalyst or templating agent to a well-established process. In the case of the phthalocyanines it was the chance observation that the iron vats used to prepare phthalimide from phthalic anhydride and ammonia contained small amounts of an insoluble blue residue at the end of the process [1–3]. The discovery was made by Dandridge in 1928 at the Grangemouth works of Scottish Dyes Ltd, and the compound was later shown to be the iron complex of phthalocyanine. A patent for the process was applied for in 1928 and granted the following year [4]. Quick to capitalize on this breakthrough the company found that the copper phthalocyanine complex gave a more stable blue dye and marketed it in 1935 under the name Monastal Blue. The attraction of phthalocyanine complexes of transition metals lies in their bright colours and thermal stabilities. Many can be heated to 400 °C under vacuum, subliming without degradation, and the copper complex is stable to 900 °C. To date phthalocyanines have been shown to complex alkali metals, alkaline earths, metals and metalloids of the p-block, and majority of the transition metals. Indeed it appears that silver and mercury are the only transition metals not to form complexes with them [5]. From the early 1930s onwards the chemistry of the phthalocyanines was investigated by the Linstead group working at London's Imperial College, with the initial reports appearing in six consecutive papers [6]. Their work showed that the parent ligand comprised four isoindole units linked by imines, a fact soon confirmed by Robertson's X-ray studies [7]. The similarity to porphyrins was immediately apparent: all that was required was the replacement of the imine bridges with methines. Since the 1930s phthalocyanine–metal complexes have become a mainstay of the dyestuffs industry with a contemporary role to play in colour printers, corrosion inhibitors and compact discs used for data storage.

Perhaps of greater benefit is the use of phthalocyanines in photodynamic therapy [8]. This treatment utilizes the ability of phthalocyanine complexes, when irradiated, to generate cytotoxic and antiviral species, such as hydroxyl radicals and singlet oxygen, that in turn effect the tissue surrounding the complexes. The greatest use of this form of therapy is with skin-related conditions, as the incident light does not have to penetrate too far into the tissue to activate the complexes.

Phthalocyanines have been derivatized in a variety of ways. Perhaps those of greatest interest to supramolecular chemists incorporate crown ethers, particularly [15]crown-5, and therefore have binding sites for both transition and alkali metals. The combination of binding sites results in interesting colligative effects: the degree of self-association is often governed by the concentration of alkali metal present [9]. Most phthalocyanines are terameric, yet despite the clear steric constraints, it is also possible to isolate trimeric forms of phthalocyanines which go under the designation of subphthalocyanines.

The example given here is copper phthalocyanine (**13**); its synthesis is illustrated in Figure 2.3, and is one of the easiest to prepare. Many metal complexes are known but the copper complex, shown in Figure 2.4, is perhaps the most representative. The initial reaction often generates the β-form of the material which can be transformed to the more stable α-form by grinding and precipitation

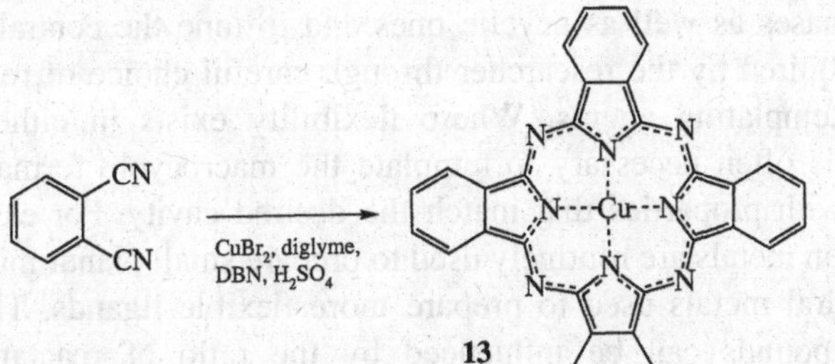

Figure 2.3 Synthesis of copper phthalocyanine (**13**)

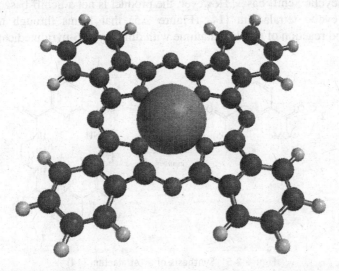

Figure 2.4 Simulated structure of the copper phthalocyanine complex

from concentrated acid. The original patented process started with phthalic anhydride which was heated with copper(II) chloride at temperatures exceeding 220 °C followed by addition of gaseous ammonia. The product was isolated using a complex work-up procedure. The method given here uses phthalonitrile which, though more effective than the earlier route, was too expensive for industrial application.

In the previous chapter, examples of Schiff base condensations have been given in which compounds form through reactions between aldehydes and amines to generate the target imines. This method introduces convergent donor atom functionality to the podands through a remarkably facile and generally high-yielding route. The cyclic porphyrins and phthalocyanines take convergence a step further by creating a rigid binding pocket for cations, particularly transition metals, which require a square planar array of donor atoms. It is possible to form cyclic Schiff bases as well as acyclic ones and to tune the central cavity to the dimensions required by the researcher through careful choice of reagents and, in some cases, templating agents. Where flexibility exists in either diamine or dialdehyde it is often necessary to template the macrocycle formation around a metal cation with properties that match the desired cavity. For example, square planar transition metals are routinely used to prepare small planar macrocycles and larger tetrahedral metals used to prepare more flexible ligands. The size of the resulting compounds can be influenced by the ratio of reactants: for small compounds it will be 2:2 but larger macrocycles can form with 3:3, 4:4 or even higher ratios.

The second synthesis in this section is a variation on the methodology required to prepare cyclic Schiff bases. However, the product is not a Schiff base *per se* but a related cyclic tetralactam (**14**) (Figure 2.5) that forms through the facile, untemplated reaction of ethylenediamine with dimethyl 2,6-pyridinedicarboxylate,

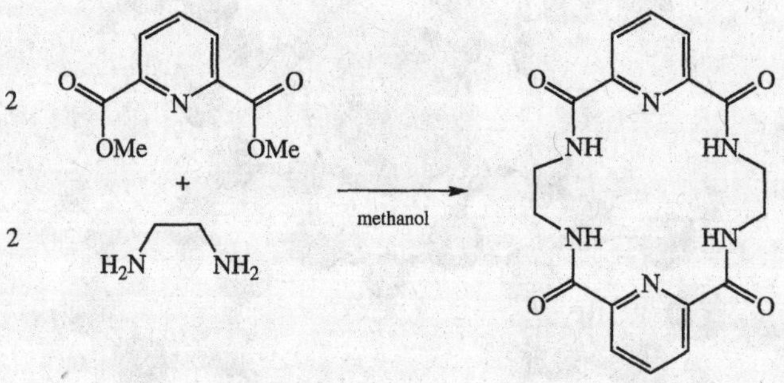

Figure 2.5   Synthesis of a tetralactam (**14**)

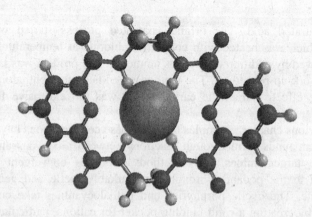

Figure 2.6   Simulated structure of a tetralactam fluoride complex

first reported by Weber and Vögtle in 1976 [10]. The tetralactam has since been shown by Jurczak to bind a number of anions such as the fluoride complex in Figure 2.6 [11].

[1] The phthalocyanines, A. B. P. Lever, *Adv. Inorg. Chem. Radiochem.*, 1965, **7**, 27.
[2] *Template synthesis of macrocyclic compounds*, N. V. Gerbeleu, V. B. Arion and J. Burgess, Wiley-VCH, Weinheim, 1999.
[3] Industrial applications of phthalocyanines, P. Gregory, *J. Porphyrins Phthalocyanines*, 2000, **4**, 432.
[4] Improvements in and relating to the manufacture and use of colouring materials, A. G. Dandridge, H. A. E. Drescher, J. Thomas and Scottish Dyes Ltd., GB Patent 322169, November 18, 1928.
[5] *Coordination Compounds of Porphyrins and Phthalocyanines*, B. D. Berezin, Wiley, New York, 1981.
[6] Phthalocyanines. Part I. A new type of synthetic colouring matters, R. P. Linstead, *J. Chem. Soc.*, 1935, 1016; Phthalocyanines. Part II. The preparation of phthalocyanine and some metallic derivatives from *o*-cyanobenzamide and phthalimide, G. T. Byrne, R. P. Linstead and A. R. Lowe, *J. Chem. Soc.*, 1935, 1017; Phthalocyanines. Part III. Preliminary experiments on the preparation of phthalocyanines from phthalonitrile, R. P. Linstead and A. R. Lowe, *J. Chem. Soc.*, 1935, 1022; Phthalocyanines. Part IV. Copper phthalocyanines, C. E. Dent and R. P. Linstead, *J. Chem. Soc.*, 1935, 1027; Phthalocyanines. Part V. The molecular weight of magnesium phthalocyanine, R. P. Linstead and A. R. Lowe, *J. Chem. Soc.*, 1935, 1031; Phthalocyanines. Part VI. The structure of the phthalocyanines, C. E. Dent, R. P. Linstead and A. R. Lowe, *J. Chem. Soc.*, 1935, 1033.
[7] An X-ray study of the structure of the phthalocyanines. Part I. The metal-free, nickel, copper, and platinum compounds, J. M. Robertson, *J. Chem. Soc.*, 1935, 615, An X-ray study of the phthalocyanines. Part II. Quantitative structure determination of the metal-free compound, J. M. Robertson, *J. Chem. Soc.*, 1936, 1195; An X-ray study of the phthalocyanines. Part III. Quantitative structure determination of nickel phthalocyanine, J. M. Robertson and I. Woodward, *J. Chem. Soc.*, 1937, 219; An X-ray study of the phthalocyanines. Part IV. Direct quantitative analysis of the platinum compound, J. M. Robertson and I. Woodward, *J. Chem. Soc.*, 1940, 36.
[8] Synthesis, photophysical and photochemical aspects of phthalocyanines for photo-dynamic therapy, A. C. Tedesco, J. C. G. Rotta and C. N. Lunardi, *Curr. Org. Chem.*, 2003, **7**, 187.
[9] Cation-induced or solvent-induced supermolecular phthalocyanine formation – crown-ether substituted phthalocyanines, N. Kobayashi and A. B. P. Lever, *J. Am. Chem. Soc.*, 1987, **109**, 7433.
[10] Ligandstruktur und Komplexierung VI. Übergangsmetallkomplexe neuer Kronenätheramine und -sulfide, E. Weber and F. Vögtle, *Justus Liebigs Ann. Chem.*, 1976, 891.
[11] A new macrocyclic polylactam-type neutral receptor for anions - structural aspects of anion recognition, A. Szumna and J. Jurczak, *Eur. J. Org. Chem.*, 2001, 4031.

Preparation of copper phthalocyanine

Copper phthalocyanine (13)

Reagents
Phthalonitrile
Copper (II) bromide (anhydrous)
1,5-Diazabicyclo[4.3.0]non-5-ene (DBN)
2-Methoxyethyl ether (diglyme)
Distilled water
Concentrated sulphuric acid [CORROSIVE]

Equipment
Round-bottomed flask (100 mL)
Reflux condenser
Heating/stirring mantle and stirrer bar
Conical flask (250 mL)
Büchner funnel and flask

Note: This reaction should be carried out in a fume hood.

Add phthalonitrile (2.6 g, 20 mmol), anhydrous copper (II) bromide (2.9 g, 13 mmol), 1,5-diazabicyclo[4.3.0]non-5-ene (2.0 g, 16 mmol) and diglyme (10 ml) to a 100 mL round-bottomed flask containing a magnetic stirrer and fit a reflux condenser. Using a heating/stirring mantle, bring the solvent to reflux (b.p. 162 °C) and continue to heat the solution for a further 2 h. Cool the dark mixture to room temperature and pour or, if solid, scrape it into a 250 mL conical flask containing distilled water (100 mL). Boil and stir the mixture to remove water-soluble impurities, then cool and add concentrated sulphuric acid (*ca.* 0.5 mL) to neutralize any remaining base. Filter the blue suspension from the green solution, dry by suction and isolate the crude product as a blue–black powder. Leave to dry further in an oven at 100 °C until a friable solid is formed. Purify by grinding the solid to a powder and dissolving it in concentrated sulphuric acid (5 mL per g product). Note that this process is highly exothermic. Leave for 30 min then slowly pour the solution into a beaker (250 mL) containing crushed ice (100 g). Rinse the solids into the beaker with distilled water (*ca.* 50 mL). Leave to stand overnight, decant any material floating on the surface, filter the denser solids using a glass fritted funnel and wash with distilled water until the washes become pale in colour (*ca.* 3 × 50 mL). Stir constantly to avoid clogging the filter. Finally wash with acetone (*ca.* 30 mL) to leave copper phthalocyanine (**13**) as a dark blue powder.

Yield: 2.8 g (quantitative); m.p.: >250 °C; IR (v, cm^{-1}): 3410, 1635, 1420, 1090; UV-vis: 680 nm.

Preparation of a cyclic tetralactam

Tetralactam (14)

Reagents
Dimethyl 2,6-pyridinedicarboxylate
Ethylenediamine [FLAMMABLE]
Methanol [FLAMMABLE; TOXIC]

Equipment
Conical flask (250 mL)
Büchner flask
Fritted filter funnel

Note: This reaction should be carried out in a fume hood.

Dissolve ethylenediamine (0.67 mL, 0.6 g, 10 mmol) in methanol (50 mL). Prepare a solution of dimethyl 2,6-pyridinedicarboxylate (1.95 g, 10 mmol) in methanol (50 mL). A little gentle warming may be necessary for complete dissolution. Combine the solutions and agitate briefly to ensure complete mixing then leave the colourless solution at room temperature for 3 to 4 days, or until no further precipitation of colourless crystals is observed. Filter to isolate the white powdery tetralactam (14), as a methanol solvate.

Yield: 0.80 g, (*ca.* 40%); m.p.: >250 °C; IR (v, cm^{-1}): 3275, 2940, 1665, 1540, 1450, 1380, 1175, 1085, 1015, 915, 850, 740, 685, 650; ^{1}H (δ, ppm; DMSO-d_6) 9.5 (br s, 4 H, ArH), 8.2 (m, 6 H, ArH), 3.6 (br s, 8 H, NCH_2CH_2N). Note: co-crystallized methanol is seen at 4.1 (br s, OH) and 3.2 (s, 3H, CH_3) ppm.

2.3 Serendipitous Macrocycles

In 1967 Charles J. Pedersen, then 2 years away from retirement as a research chemist with Du Pont, published an extensive paper detailing the preparation of 33 cyclic polyethers [1]. His thorough investigation of these macrocycles was the culmination of several years' work initiated by the accidental preparation of dibenzo[18]crown-6 [2]. The compound formed as a result of trace amounts of catechol that had contaminated 2-(hydroxyphenoxy)tetrahydropyran used in the synthesis of the original target compound, bis[2-(hydroxyphenoxy)ethyl]ether. The dibenzocrown ether thus became the first of the crown ethers, as Pedersen called them, to be isolated.

Whilst the yields of crown compounds incorporating benzene or cyclohexane are in the 20 to 80 per cent range, Pedersen only obtained [18]crown-6 in 1.8 per cent yield. As this compound found widespread application, notably as a phase transfer catalyst, a better synthesis was required for aspiring supramolecular chemists. Fortunately Greene reported superior conditions for the formation of crown ethers and described the preparation of [18]crown-6, [21]crown-7, [24]crown-8 and aza[18]crown-6 in 1972 [3]. The synthesis of [18]crown-6 (15) described here, and illustrated in Figure 2.7, is based on Greene's method where potassium deprotonates a polyether which then forms a template for reaction with a ditosylated polyether. The crowns can complex a wide variety of metals that exhibit spherical symmetry, including many lanthanides, as shown in Figure 2.8.

As well as spherical cations, [18]crown-6 is also predisposed to bind protonated species with trigonal symmetry such as H_3O^+, as will be shown in Chapter 5, and terminal ammonium groups. However, in order to harness the binding information it is necessary to build some form of signal transducer, a fluorescent group for example, into the crown structure so that the binding event can be signalled. This is best done through derivatives such as benzocrowns and azacrown lariat ethers. The latter will be described in Section 2.5.

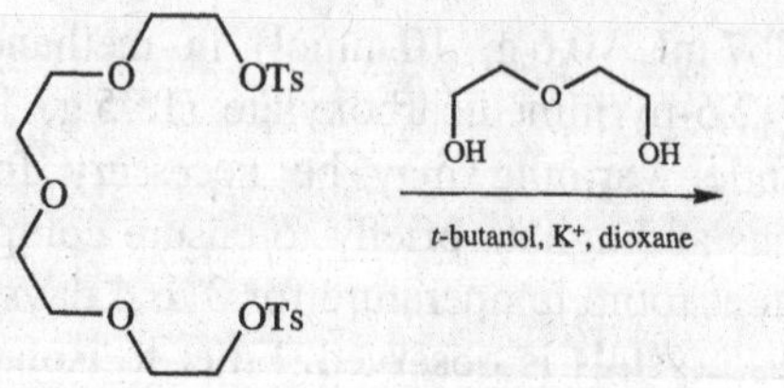

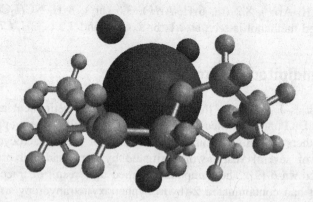

Figure 2.7 Synthesis of [18]crown-5 (**15**)

Figure 2.8 Simulated structure of a lanthanum [18]crown-6 complex

The literature on crown ethers and their complexes is vast, with approximately 5000 papers citing the term as a key word, and the best introduction to their chemistry is through review articles and books [4–7]. Suffice to say that applications run the range from simple cation detection by piezoelectric sensors [8] to the detection of metals in blood [9] and the more complicated problems of neurotransmitter [10] and biologically relevant sugar derivatives [11].

[1] Cyclic polyethers and their complexes with metal salts, C. J. Pedersen, *J. Am. Chem. Soc.*, 1967, **89**, 7017.

[2] Chemical crowns and crypts, in *Serendipity: Accidental Discoveries in Science*, R. M. Roberts, J. Wiley & Sons, Inc., New York, 1989.

[3] 18-Crown-6: a strong complexing agent for alkali metal cations, R. N. Greene, *Tetrahedron Lett.*, 1972, **13**, 1793.

[4] To appreciate the wealth and diversity of crown ethers and their inclusion possibilities see for example: *Macrocyclic Polyether Synthesis*, G. W. Gokel and S. H. Korzeniowski, vol. 13 in *Reactivity and structure concepts in organic chemistry*, Springer-Verlag, Berlin, 1982.

[5] Thermodynamic and kinetic data for macrocycle interaction with cations and anions, R. M. Izatt, K. Pawlak, J. S. Bradshaw and R. L. Bruening, *Chem. Rev.*, 1991, **91**, 1721.

[6] Thermodynamic and kinetic data for macrocycle interaction with cations, anions, and neutral molecules, R. N. Izatt, K. Pawlak, J. S. Bradshaw and R. L. Bruening, *Chem. Rev.*, 1995, **95**, 2529.

[7] *Crown ethers and cryptands* (*Monographs in Supramolecular Chemistry*), G. W. Gokel, RSC, Cambridge, 1991.

[8] The quantification of potassium using a quartz crystal microbalance, M. Teresa, S. R. Gomes, K. S. Tavares and J. A. B. P. Oliveira, *Analyst*, 2000, **125**, 1983.

[9] A novel optically based chemosensor for the detection of blood Na^+, T. Gunnlaugsson, M. Nieuwenhuyzen, L. Richard and V. Thoss, *Tetrahedron Lett.*, 2001, **42**, 4725.

[10] Fluorescent signalling of the brain neurotransmitter γ-aminobutyric acid and related amino acid zwitterions, A. P. de Silva, H. Q. Nimal Gunaratne, C. McVeigh, G. E. M. Maguire, P. R. S. Maxwell and E. O'Hanlon, *Chem. Commun.*, 1996, 2191.

[11] Selective D-glucosamine hydrochloride fluorescence signalling based on ammonium cation and diol recognition, C. R. Cooper and T. D. James, *Chem. Commun.*, 1997, 1419.

Preparation of crown ethers

[18]Crown-6 (15)

Reagents
Tetraethylene glycol ditosylate (3)
t-Butanol [FLAMMABLE]
Potassium *t*-butoxide [CORROSIVE; FLAMMABLE]
Diethylene glycol [FLAMMABLE]
Dry 1,4-dioxane [FLAMMABLE]
Dichloromethane [TOXIC]
Hexane [FLAMMABLE; NEUROTOXIN]
Distilled water
Acetonitrile [FLAMMABLE]

Equipment
Three-necked round-bottomed flask (1 L)
Pressure equalized addition funnel
Thermometer
Inert atmosphere line
Rotary evaporator
Glassware for workup

Note: This procedure should be carried out, wherever practical, in a fume hood as a large volume of flammable *t*-butanol is used.

Under an inert atmosphere, prepare a solution of *t*-butanol (350 mL), potassium *t*-butoxide (32.5 g, 0.29 mol) and diethylene glycol (22.8 mL, 25.5 mL, 0.24 mol).

Stir at 40 °C in a 1 L three-necked round-bottomed flask, fitted with a pressure-equalized addition funnel and a thermometer, until a clear solution is obtained. Add a solution of tetraethylene glycol ditosylate, **3**, (49 mL, 60.5 g, 0.12 mol) in dry 1,4-dioxane (180 mL) dropwise over 2 h from the addition funnel. Once all the ditosylate has been added continue to stir for a further 2 h and allow the solution to cool to room temperature.

Filter the precipitated sodium tosylate and wash with dichloromethane (2 × 50 mL or until the washes are colourless) to give an orange solution. Dry the filtrate over magnesium sulphate, filter and remove volatiles on the rotary evaporator to leave a brown viscous mass. Add distilled water (50 mL) and wash with hexane (50 mL) to remove acyclic by-products. Extract the red–brown aqueous solution with dichloromethane (100 mL then 2 × 50 mL), combine the organic phases and wash with distilled water (100 mL). Remove dichloromethane on the rotary evaporator, dissolve in acetonitrile (25 mL) and leave in the coldest freezer available. The [18]crown-6-acetonitrile complex forms over several days in a freezer and may be isolated by filtration. The acetonitrile is removed by gentle heating under high vacuum to leave the product as a colourless oil which rapidly recrystallizes. [18]Crown-6 (**15**), is obtained as a white hydroscopic solid.

Yield: 9.6 g, (30%); m.p.: 42–45 °C; IR (v, cm^{-1}): 2940, 1970, 1670, 1355, 1190, 1105, 960, 835, 680, 560; ^{1}H (δ, ppm; CDCl$_3$) 3.6 (s, 10 H, CH_2CH_2O).

2.4 Adding Functionality to the Crowns

Once the synthesis of crown ethers had been established the next development in the field was to incorporate donor groups other than oxygen. A change in donor groups required a new synthetic route, first achieved by Lehn in 1969 just two years after Pedersen's groundbreaking work, in which diaza[18]crown-6 was prepared as a precursor to the encapsulating 2,2,2-cryptand [1]. The first monoazacrown ether, aza[18]crown-6, was prepared by Greene [2]; however, there have been many variations and improvements since the original synthesis was reported [3–5]. Azacrown ethers have found a variety of applications usually related to their ability to bind cations, such as lanthanum (Figure 2.9) [6], or their subsequent functionalization to produce N-azacrown lariat ethers as will be seen in the following section.

Two examples are included here, aza[15]crown-5 (**16**) and aza[18]crown-6 (**17**), prepared in a similar manner to the analogous all-oxygen crowns. Both reactions are templated around the appropriately sized alkali metal, as shown

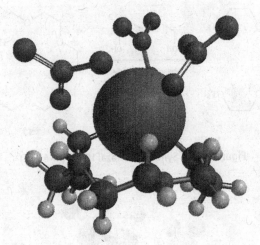

Figure 2.9 Simulated structure of [La·**16**] (NO$_3$)$_3$

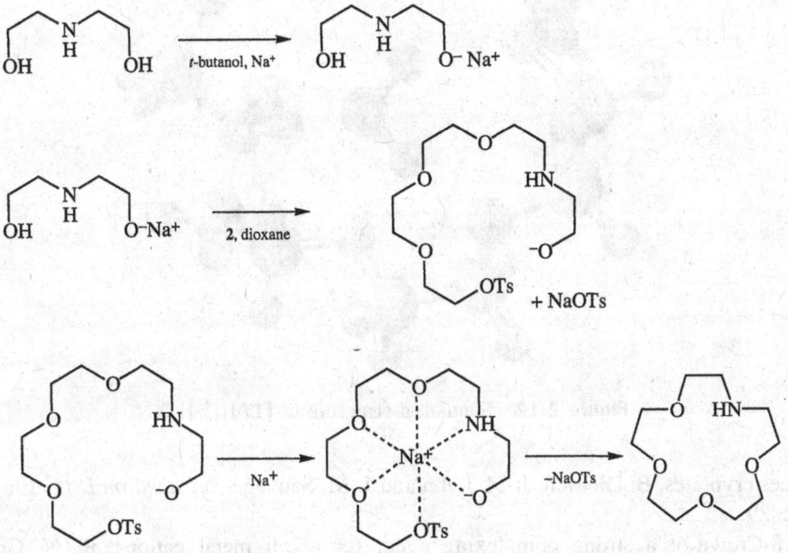

Figure 2.10 Synthesis of aza[15]crown-5 (**16**) illustrating the templating effect of sodium

in Figures 2.10 and 2.11. Note that, as in the case of [18]crown-6, ratios of reagents are not stoichiometric but have been optimized to give the best yields [5]. The products are readily protonated but, unlike [18]crown-6, do not stabilize the hydronium ion as shown by the simulation in Figure 2.12.

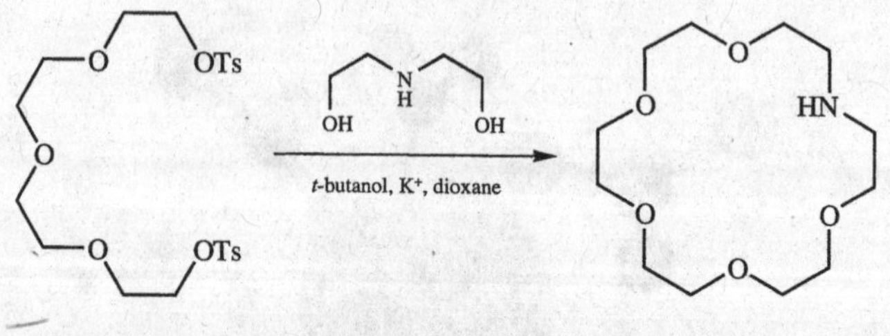

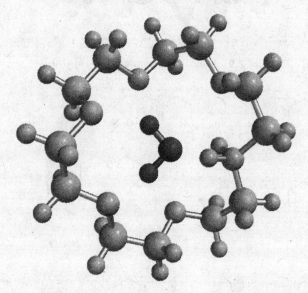

Figure 2.11 Synthesis of aza[18]crown-6 (**17**)

Figure 2.12 Simulated structure of $[17 \cdot H]^+ \cdot H_2O$

[1] Les cryptates, B. Dietrich, J.-M. Lehn and J.-M. Sauvage, *Tetrahedron Lett.*, 1969, **10**, 2889.

[2] 18-Crown-6: a strong complexing agent for alkali metal cations, R. N. Greene, *Tetrahedron Lett.*, 1972, **13**, 1793.

[3] The formation of complexes between aza derivatives of crown ethers and primary alkyl ammonium salts. Part 1. Monoaza derivatives, M. R. Johnson, I. O. Sutherland and R. F. Newton, *J. Chem. Soc., Perkin Trans. 1*, 1979, 357.

[4] Crown–cation complex effects III. Chemistry and complexes of monoaza-18-crown-6, G. W. Gokel and B. J. Garcia, *Tetrahedron Lett.*, 1977, **18**, 317.

[5] Synthesis of monoaza crown ethers from *N,N*-di[oligo(oxyalkylene)]amines and oligoethylene glycol di(*p*-toluenesulfonates) or corresponding dichlorides, H. Maeda, S. Furuyoshi, Y. Nakatsuji and M. Okahara, *Bull. Chem. Soc. Jpn.*, 1983, **56**, 212.

[6] Lanthanide and actinide complexes of monoaza-15-crown-5. Synthesis and crystal structures of [La(monoaza-15-crown-5)(NO₃)₃] and [UO₂(NO₃)₂]₂(μ-H₂O)₂(monoaza-15-crown-5)₂, P. J. Cragg, S. G. Bott and J. L. Atwood, *Lanth. Act. Res.*, 1988, **2**, 265.

Preparation of azacrown ethers

Aza[15]crown-5 (16)

Reagents

Triethylene glycol ditosylate (**2**)
t-Butanol [FLAMMABLE]
Sodium *t*-butoxide [CORROSIVE; FLAMMABLE]
Diethanolamine [HARMFUL]
Dry 1,4-dioxane [FLAMMABLE]
Dichloromethane [TOXIC]
Distilled water
Anhydrous magnesium sulphate
Diethyl ether [FLAMMABLE]

Equipment

Three-necked round-bottomed flask (1 L)
Pressure-equalized addition funnel
Thermometer
Inert atmosphere line
Rotary evaporator
Vacuum line
Kugelrohr

Note: This procedure should be carried out, wherever practical, in a fume hood as a large volume of flammable *t*-butanol is used.

Under an inert atmosphere, prepare a solution of *t*-butanol (250 mL), sodium *t*-butoxide (20.0 g, 0.21 mol) and diethanolamine (17.0 mL, 18.9 g, 0.09 mol). Stir at 40 °C in a 1 L three-necked round-bottomed flask, fitted with a pressure-equalized addition funnel and a thermometer, for 30 min. Add a solution of triethylene glycol ditosylate, **2**, (40.0 g, 0.9 mol) in dry 1,4-dioxane (150 mL) dropwise over 2 h from the addition funnel. During the addition the reaction appears to be slightly exothermic and holds 40 °C with minimal external heating. Once all the ditosylate has been added continue to stir for a further 1 h and allow the solution to cool to room temperature.

Filter the precipitated sodium tosylate and wash with dichloromethane until the washes are colourless (*ca.* 100 mL). Remove volatiles on the rotary evaporator, add distilled water (50 mL) and wash with hexane (50 mL) to remove acyclic by-products. Extract the pale yellow aqueous solution with dichloromethane (2 × 100 mL, 1 × 50 mL), combine the organic phases and wash with distilled water (50 mL). Dry the dichloromethane extract with anhydrous magnesium sulphate using enough to clarify the solution (1 to 2 g), filter and remove solvent on the rotary evaporator. Purify by Kugelrohr distillation (b.p 110–115 °C, 0.1 mmHg) to yield the product, aza[15]crown-5 (**16**), as a white hydroscopic solid.

Yield: 11 g (55%); m.p.: 35–37 °C; IR (v, cm^{-1}): 3320, 2940, 2860, 1460, 1350, 1120; ^1H (δ, ppm; CDCl$_3$) 3.6 (m, 16 H, CH_2CH_2O), 2.7 (t, 4 H, NHCH_2), 2.6 (s, 1 H, NHCH$_2$).

Aza[18]crown-6 (17)

Reagents	**Equipment**
Tetraethylene glycol ditosylate (3)	Three-necked round-bottomed flask (1 L)
t-Butanol [FLAMMABLE]	Pressure-equalized addition funnel
Potassium t-butoxide [CORROSIVE;	Thermometer
FLAMMABLE]	Inert atmosphere line
Diethanolamine [HARMFUL]	Rotary evaporator
Dry 1,4-dioxane [FLAMMABLE]	Vacuum line
Dichloromethane	Kugelrohr
Distilled water	

Note: This procedure should be carried out, wherever practical, in a fume hood as a large volume of flammable t-butanol is used.

Under an inert atmosphere, prepare a solution of t-butanol (350 mL), potassium t-butoxide (32.5 g, 0.29 mol) and diethanolamine (23.0 mL, 25.2 g, 0.12 mol). Stir at 40 °C in a three-necked round-bottomed flask (1 L), fitted with a pressure-equalized addition funnel and a thermometer, until a clear solution is obtained. Add a solution of tetraethylene glycol ditosylate, **3**, (48.5 mL, 60.5 g, 0.12 mol) in dry 1,4-dioxane (180 mL) dropwise over 2 h from the addition funnel. As with the previous example, this reaction appears to hold at 40 °C during the addition. Once all the ditosylate has been added continue to stir for a further 1 h and allow the solution to cool to room temperature.

Filter the precipitated potassium tosylate and wash with dichloromethane until the washes are colourless (usually 2 × 100 mL). Remove volatiles on the rotary evaporator, add distilled water (100 mL) and wash with hexane (50 mL) to remove acyclic by-products. Extract the yellow aqueous solution with dichloromethane (2 × 100 mL then 1 × 50 mL), combine the organic phases and wash with distilled water (3 × 100 mL). Remove dichloromethane on the rotary evaporator and isolate aza[18]crown-6 (**17**) by Kugelrohr distillation (b.p 95–100 °C, 0.2 mmHg) as a white crystalline solid.

Yield: 15 g (45%); m.p.: 48–50 °C; IR (v, cm^{-1}): 3300, 2940, 2850, 1460, 1350, 1120; ^1H (δ, ppm; CDCl$_3$) 3.6 (m, 20 H, CH_2CH_2O), 2.8 (t, 4 H, NHCH_2), 2.7 (s, 1 H, NHCH$_2$).

2.5 Azacrowns with Sidearms

It is possible to increase the binding capability of a crown ether through the introduction of one or more sidearms to form a lariat ether. While a carbon atom in the crown can be used as a point of attachment, it is far easier to modify an azacrown through addition of a substituent to the nitrogen atom. This may be undertaken prior to cyclization, by replacing diethanolamine in the preparation of **16** or **17** with a suitably N-functionalized derivative [1], or in a separate step following cyclization [2,3]. The latter approach is useful if a range of lariat ethers containing different sidearms is required, though the former may be preferred if a large quantity of a single derivative is desired. It is possible to follow a variety of synthetic routes to aza-, diaza- and triazacrown ethers [4,5] and to use these compounds for a variety of purposes from micelle formation [6] and artificial transmembrane ion transport [7] to setting cement [8]. Most recently these compounds have helped to model biologically important phenomena such as cation-π interactions with amino acids [9].

The examples given here, N-allylaza[15]crown-5 (**18**) and N-cinnamyla-za[18]crown-6 (**19**), are typical of a range of lariat ethers in which the sidearm contains an allyl moiety. This leads to some unexpected complexation modes that are dependent upon the conformation of the sidearm. Complexes of potassium with N-functionalized aza[18]crown-6 derivatives do not involve sidearm participation in supramolecule formation [1] whereas potassium is encapsulated by the aza[15]crown-5 analogues [10]. N-Allylaza[15]-crown-5 is of interest as different models for self-assembly can be engineered depending on the cations and anions incorporated. With potassium hexafluorophosphate a complex involving two lariat ethers, two perching cations and two metal-bound anions crystallizes from methanol, as shown in Figure 2.13. In the structure the two ligand-cation-anion assemblies are related through C_2 symmetry with the allyl sidearms almost perpendicular to the plane of the crown oxygen atoms. If the same ligand is

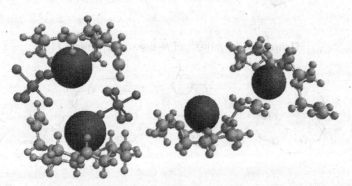

Figure 2.13 Structures of $\{[K \cdot 18]^+ \ PF_6^-\}_2$ (left) and the $\{[Ag \cdot 18]^+\}_n$ motif (right)

treated with silver hexafluoroantiminate it adopts a conformation in which the silver cation sits slightly above the plane of the crown ether allowing the cation to form a π-interaction with the allyl sidearm of an adjacent ligand to give a linear polymer (Figure 2.13). When the analogous *N*-cinnamylaza[15]-crown-5 lariat ether is treated with potassium hexafluorophosphate a similar dimeric ligand–cation–anion complex forms; yet when treated with sodium iodide in methanol dimerization does not occur as the smaller cation fits within the macrocyclic cavity (Figure 2.14). Here, the ligand adopts the geometry seen with *N*-allylaza[15]-crown-5 silver complexes: the sidearm is almost coplanar with the crown oxygen atoms.

The syntheses of both compounds are straightforward, as seen from the schemes in Figures 2.15 and 2.16, and the general route is widely applicable. Note that if sidearms are to be introduced using an alkyl halide without a double bond in the

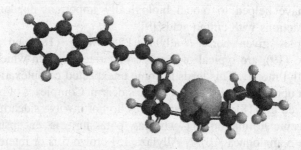

Figure 2.14 Simulated structure of the *N*-cinnamylaza[15]crown-5 complex with NaI

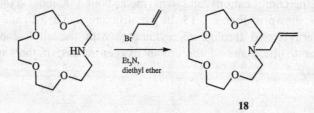

18

Figure 2.15 Synthesis of *N*-allylaza[15]crown-5 (**18**)

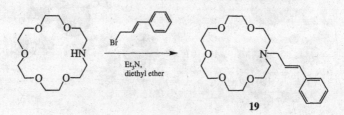

19

Figure 2.16 Synthesis of *N*-cinnamylaza[18]crown-6 (**19**)

β-position, the reaction will take significantly longer and may require a base such as potassium carbonate in place of triethylamine.

[1] 12-, 15-, and 18-Membered-ring nitrogen-pivot lariat ethers: syntheses, properties, and sodium and ammonium cation binding properties, R. A. Schultz, B. D. White, D. M. Dishong, K. A. Arnold and G. W. Gokel, *J. Am. Chem. Soc.*, 1985, **107**, 6659.

[2] Formation of an organometallic coordination polymer from the reaction of silver(I) with a non-complementary lariat ether, P. D. Prince, P. J. Cragg and J. W. Steed, *Chem. Commun.*, 1999, 1179.

[3] Organometallic crown ethers 1. Metal acyl binding to a crown ether held cation, S. J. McLain, *J. Am. Chem. Soc.*, 1983, **105**, 6355.

[4] *N,N'*-Bis(disubstituted)-4,13-diaza-18-crown-6 derivatives having pi-donor-group-sidearms: correlation of thermodynamics and solid state structures, K. A. Arnold, A. M. Viscariello, M. Kim, R. D. Gandour, F. R. Fronczek and G. W. Gokel, *Tetrahedron Lett.*, 1988, **29**, 3025.

[5] Tribracchial lariat ethers, 'TriBLEs', based on 4, 10, 16-triaza-18-crown-6: an apparent limit to sidearm contributions in lariat ether molecules, S. R. Miller, T. P. Cleary, J. E. Trafton, C. Smeraglia, F. R. Fronczek and G. W. Gokel, *J. Chem. Soc., Chem. Commun.*, 1989, 806.

[6] Azacrown ethers as amphiphile headgroups: formation of stable aggregates from two- and three-armed lariat ethers, S. L. De Wall, K. Wang, D. R. Berger, S. Watanabe, J. C. Hernandez and G. W. Gokel, *J. Org. Chem.*, 1997, **62**, 6784.

[7] Towards a redox-active artificial ion channel, C. D. Hall, G. J. Kirkovits and A. C. Hall, *Chem. Commun.*, 1999, 1897.

[8] Design and synthesis of macrocyclic ligands for specific interaction with crystalline ettringite and demonstration of a viable mechanism for the setting of cement, J. L. W. Griffin, P. V. Coveney, A. Whiting and R. Davey, *J. Chem. Soc., Perkin Trans. 2*, 1999, 1973.

[9] Sodium cation complexation behavior of the heteroaromatic sidechains of histidine and tryptophan, J. Hu, L. J. Barbour, R. Ferdani and G. W. Gokel, *Chem. Commun.*, 2002, 1810.

[10] *C*-Donor lariat ether 'scorpionates', P. Arya, A. Channa, P. J. Cragg, P. D. Prince and J. W. Steed, *New J. Chem.*, 2002, **26**, 440.

Preparation of azacrown lariat ethers

N-Allylaza[15]crown-5 (18)

Reagents
Allyl bromide [LACHRYMATOR]
Triethylamine [FLAMMABLE]
Aza[15]crown-5, **16**
Dry diethyl ether [FLAMMABLE]

Equipment
Two-necked round-bottomed flask (100 mL)
Magnetic stirrer and stirrer bar
Inert atmosphere line
Rotary evaporator

Distilled water
Dichloromethane [TOXIC]

Vacuum line
Kugelrohr
Glassware for work-up

Note: Work in a fume hood as allyl bromide is a powerful lachrymator.

Prepare a solution of allyl bromide (1.0 mL, 1.4 g, 12 mmol) in dry diethyl ether (10 mL) and add it dropwise to a stirred solution of aza[15]crown-5, **16**, (2.6 g, 12 mmol) and triethylamine (1.8 mL, 1.3 g, 13 mmol) in dry diethyl ether (25 mL) at room temperature under an inert atmosphere. Triethylamine hydrochloride precipitation occurs almost immediately but it is worth stirring the reaction mixture for several hours more to ensure complete reaction. Remove the solvent by rotary evaporation and add distilled water (25 mL) to dissolve the precipitate. Extract with dichloromethane (3 × 25 mL) and remove the organic solvent by rotary evaporation. Distil the resulting pale green oil by Kugelrohr (b.p. 118–120 °C, 0.25 mmHg) to yield *N*-allylaza[15]crown-5 (**18**) as a colourless oil.

Yield: 2.5 g (80%); IR (v, cm^{-1}): 3090, 2950, 2880, 1650, 1480, 1460, 1425, 1300, 1260, 1130, 1000, 940, 845; ^{1}H (δ, ppm; CDCl$_3$) 5.7 (m, 1 H, C*H*CH$_2$), 5.0 (dd, 2 H, CHC*H*$_2$), 3.6 (m, 16 H, C*H*$_2$C*H*$_2$O), 2.9 (d, 2 H, NC*H*$_2$CHCH$_2$), 2.6 (t, 4 H, NC*H*$_2$CH$_2$O).

N-Cinnamylaza[18]crown-6 (19)

Reagents
Cinnamyl bromide
Triethylamine [FLAMMABLE]
Aza[18]crown-6, **17**
Dry diethyl ether [FLAMMABLE]
Distilled water
Aqueous potassium carbonate

Equipment
Round-bottomed flask (100 mL)
Magnetic stirrer and stirrer bar
Calcium chloride guard tube
Glassware for workup
Rotary evaporator

Note: Work in a fume hood or well-ventilated laboratory.

In a 100 mL round-bottomed flask, prepare a solution of aza[18]crown-6, **17**, (3.6 g, 15.2 mmol) and triethylamine (4.2 mL, 3.0 g, 30 mmol) in dry diethyl ether (25 mL) at room temperature. Dissolve cinnamyl bromide (3.0 g, 15.2 mmol) in dry diethyl ether (30 mL), add it slowly to the stirred azacrown solution and fit a calcium chloride guard tube (or alternatively run the reaction under an inert atmosphere). Triethylamine hydrobromide precipitation occurs almost immediately; however, it is worth stirring the mixture for several hours more to ensure complete reaction. Remove 50 per cent of the solvent by rotary evaporation, wash with aqueous potassium carbonate (25 mL, 2 M solution) and dry over magnesium sulphate. Filter and remove the organic solvent by rotary evaporation. The product, N-cinnamylaza[18]crown-6 (**19**), is isolated as a yellow oil.

Yield: 1.1 g, (60%); IR (v, cm^{-1}): 3080, 3040, 2980, 1610, 1595, 1460, 1360, 1300, 1270, 1120, 950, 800, 745, 695; ^{1}H (δ, ppm; CDCl$_3$) 7.4 (m, 2 H, ArH), 7.3 (m, 2 H, ArH), 7.1 (m, 1 H, ArH), 6.5 (d, 1H, ArCH=), 6.3 (m, 1H, CH$_2$CH=), 3.6 (m, 20 H, CH_2CH$_2$O), 3.4 (d, 2 H, NCH_2CH=), 2.7 (t, 4 H, NCH$_2$CH$_2$O).

2.6 Water-Soluble Macrocycles

For macrocycles to have biological or medicinal relevance it is importance that they are water soluble or at least amphiphilic. While this property is true of some macrocycles, such as the sulphonate derivatives of calixarenes, it is most evident in the cyclic sugars, the cyclodextrins. These compounds are of great importance in supramolecular chemistry: both a volume of *Comprehensive Supramolecular Chemistry* [1] and an entire edition of *Chemical Reviews* [2] are devoted to them.

Cyclodextrins were originally observed by Villiers, who reported in 1891 on crystalline products that formed in less than 1 per cent yield following the degradation of starch by *Bacillus amylobacter* (or possibly the presence of trace *Bacillus macerans*). In 1904 Schardinger reported that acetone and ethanol could be obtained from mixtures of starch-containing plants and sugar in the presence of *Bacillus macerans*. Crystalline by-products were obtained in 30 per cent yield and were later identified as cyclodextrins containing six (α), seven (β) and eight (γ) D-glucopyranoside units linked by 1,4-glycosidic bonds. The family of cyclodextrins has since increased to include a five-membered 'pre-α-cyclodextrin' and larger homologues up to and including the 12-membered ξ-cyclodextrin. It is now known that the enzyme cyclodextrin glucanotransferase catalyses the cyclization and the compounds, particularly β-cyclodextrin, are produced on an industrial scale [3].

The industrial importance of cyclodextrins was identified in 1948 when the crystal structure of γ-cyclodextrin was solved by Freudenberg and Cramer [4]. The structure revealed the cyclic nature of the compounds and their potential to include guest molecules. They have a slightly wider upper rim, with one secondary hydroxyl group per dextrin unit, than lower rim which bristles with two primary hydroxyl groups per dextrin repeat unit. Their outward pointing hydroxyl groups makes the cyclodextrins water soluble while retaining a hydrophobic central cavity. Interestingly the aqueous solubility of β-cyclodextrin at room temperature is much lower (18 g L^{-1}) than the α- or γ-forms (145 and 230 g L^{-1} respectively). All the cyclodextrins are approximately 7.8 Å in depth, however, the outer diameters and cavity sizes increase with the number of dextrin units. This gives α-cyclodextrin an overall diameter of 14.6 Å (1.46 nm) and a cavity size around 5 Å (0.5 nm), β-cyclodextrin, 15.4 Å (1.54 nm) and 6.5 Å (0.65 nm), respectively

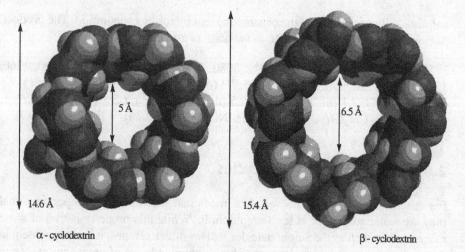

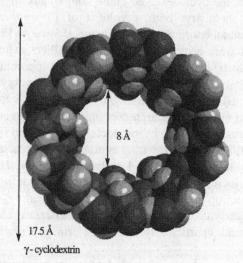

Figure 2.17 Cyclodextrins and their cavities

and γ-cyclodextrin, 17.5 Å (1.75 nm) and 8 Å (0.8 nm), respectively, as illustrated in Figure 2.17. The hydroxyl substituents on the upper and lower rims of the cyclodextrins are obvious sites for substitution. Given the chiral nature of the parent compounds, and the large number of repeat units, it is not surprising that the derivatization of cyclodextrins is a complicated matter.

Cyclodextrins appear to be extremely biocompatible. They are unable to significantly permeate biological membranes and are not absorbed by the gastro-intestinal tract yet do not induce a response from the body's immune system. Tests on rats indicate that LD_{50} values for the most toxic parent compound, β-cyclodextrin, are greater than 5 g per kg of body weight when administered orally and 0.5 g per kg of body weight when administered intravenously. This low level of toxicity, coupled to the cyclodextrins' ability to form inclusion complexes, makes them ideal excipients in pharmaceutical formulations.

A wide variety of guests, from volatile gases to steroids, can be accommodated by cyclodextrins though they are necessarily related to the cavity size. When crystalline the complexes tend to form stacked channels, face-to-face dimers, cages or layers depending on the size of the included guest molecule [5]. Although many methods are used to encapsulate guests, the simplest is to add the guest to a stirred solution or slurry of the cyclodextrin, typically in water. Once the inclusion complex has been isolated, usually by precipitation followed by drying, it is the dissociation of the host–guest complexes that becomes the key factor in the application of cyclodextrins. Ideally, complexes are stable in the dry form but dissolve readily in water prior to the guest leaving. In the case of volatile guests, however, ambient heat may be all that is required. In an example where a non-volatile guest, the antimicrobial triclosan, is included the abrasive effect of brushing teeth frees the toothpaste additive and improves the drug's efficacy threefold. Cyclo-dextrins can also be used to trap malodorous chemical species such as low molecular weight organosulphur compounds and therefore have applications in deodorants, disposable nappies and the like. The affinity of β-cyclodextrin for cholesterol also makes it ideal for removing the substance from a variety of foodstuffs to produce 'low cholesterol' versions with typically 80 per cent less cholesterol than usual.

Two major fields in which cyclodextrins have been used as agents of controlled release are pharmaceuticals and agrochemicals. In the former, they are seen as excellent drug delivery systems for hydrophobic pharmaceuticals that cannot be administered directly due to their low solubility in aqueous media. The complexes themselves are often soluble in hydrogels and other aqueous systems but cannot enter cell membranes. However, once the complex reaches the lipophilic exterior of the cell, the guest can transfer from the hydrophobic cyclodextrin cavity to the cell membrane. This technology is even being considered as a method to administer oligonuleotides to solid tumours. In the agrochemical field, cyclodex-trins are used as slow-release agents for encapsulated herbicides, fertilizers and the like. They can also aid in the remediation of contaminated environments by removing a range of contaminants from volatile gases to heavy metals.

Cyclodextrins can even be used to improve crop yield. Treatment of grain with cyclodextrins inhibits the production of starch-destroying amylase so that the more mature plants have a greater energy store and give up to 50 per cent higher yields.

From a supramolecular perspective, cyclodextrins offer a preformed cavity of defined size within a water-soluble macrocycle. The interaction between guest molecules, or substituents to the cyclodextrin itself, leads to some useful results. Hydrophobic fluorescent substituents will be 'self-included' by the macrocycle in aqueous solution until they are displaced by guests with even greater affinities [6]. The act of displacement will generate a fluorescent response that lends itself to sensor applications. It is also possible to use the hydrophobic core to induce rotaxane formation. This approach has been used by many groups, for example, Liu has recently used it to prepare water-soluble gold-polypseudorotaxanes that capture fullerenes such as C_{60} [7].

The synthesis given here is based on Bittman's route to monotosylated β-cyclodextrin (**20**) [8] which in turn provides a basis for many monosubstituted cyclodextrin derivatives. Formation of the product is easy to determine because, despite the complex ^{1}H NMR spectrum that results, the signals due to the tosylate group are very distinct. Another useful derivative, though not included here, is the aldehyde which can be prepared directly from β-cyclodextrin or its monotosylated derivative [9,10], as illustrated in Figure 2.18.

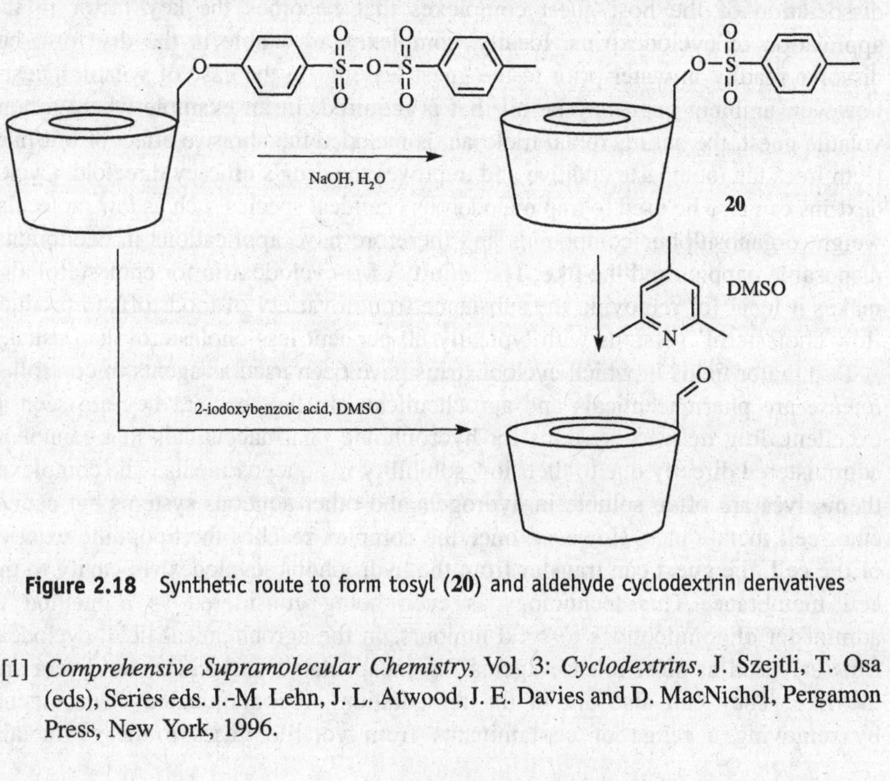

Figure 2.18 Synthetic route to form tosyl (**20**) and aldehyde cyclodextrin derivatives

[1] *Comprehensive Supramolecular Chemistry*, Vol. 3: *Cyclodextrins*, J. Szejtli, T. Osa (eds), Series eds. J.-M. Lehn, J. L. Atwood, J. E. Davies and D. MacNichol, Pergamon Press, New York, 1996.

[2] The entire July/August volume of *Chem. Rev.*, 1998, **98** is devoted to cyclodextrins.

[3] Cyclodextrins and their uses: a review, E. M. M. Del Valle, *Process Biochem.*, 2004, **39**, 1033.

[4] Die Konstitution der Schardinger-Dextrine α, β und γ, K. Freudenberg and F. Cramer, *Naturforsch.*, 1948, **3b**, 464.

[5] *Supramolecular Chemistry*, J. W. Steed and J. L. Atwood, John Wiley & Sons, Ltd. Chichester, 2000, pp. 321–334.

[6] A modified cyclodextrin with a fully encapsulated dansyl group: self inclusion in the solid state and in solution, R. Corradini, A. Dossena, R. Marchelli, A. Panagia, G. Sartor, M. Saviano, A. Lombardi and V. Pavone, *Chem. Eur. J.*, 1996, **2**, 373.

[7] Supramolecular aggregates constructed from gold nanoparticles and L-Try-CD poly-pseudorotaxanes as captors for fullerenes, Y. Liu, H. Wang, Y. Chen, C.-F. Ke and M. Liu, *J. Am. Chem. Soc.*, 2005, **127**, 657.

[8] An improved synthesis of 6-*O*-monotosyl-6-deoxy-β-cyclodextrin, N. Zhong, H.-S. Byun and R. Bittman, *Tetrahedron Lett.*, 1998, **39**, 2919.

[9] A facile synthesis of β-cyclodextrin monoaldehyde, J. Hu, C. F. Ye, Y. D. Zhao, J. B. Chang and R. Y. Guo, *Chin. Chem. Lett.*, 1999, **10**, 273.

[10] A general method for the synthesis of cyclodextrinyl aldehydes and carboxylic acids, J. Yoon, S. Hong, K. A. Martin and A. W. Czarnik, *J. Org. Chem.*, 1995, **60**, 2792.

Preparation of a monosubstituted cyclodextrin

β-Cyclodextrin monotosylate (20)

Reagents	Equipment
β-Cyclodextrin hydrate	Conical flasks (250 mL)
p-Toluenesulphonic anhydride	Magnetic stirrer and stirrer bar
[CORROSIVE]	Fritted glass filter funnel
Sodium hydroxide (2 M aqueous solution)	Büchner funnel
[CORROSIVE]	Rotary evaporator
Distilled water	
Ammonium chloride	

Note: work in a fume hood or well-ventilated laboratory.

Add p-toluenesulphonic anhydride (2.4 g, 7.5 mmol) to a 250 mL round-bottomed flask containing a vigorously stirred suspension of β-cyclodextrin monohydrate (5.7 g, 5 mmol) in distilled water (100 mL). Stir for 2 h. Add sodium hydroxide (30 mL, 2 M aqueous solution) and continue to stir for 30 min then filter the mixture through a fritted glass funnel, washing with distilled water (10 mL). Transfer the filtrate to a 250 mL conical flask and add ammonium chloride (*ca.* 3 g) to buffer the solution at pH 8. Reduce the volume of the solution to 75 mL, refrigerate overnight and isolate the product, β-cyclodextrin tosylate (**20**), by filtration washing first with distilled water (5 mL) then acetone (5 mL).

Yield: 2.85 g, (45%); m.p.: 160–162 °C; IR (v, cm^{-1}): 3270, 1650, 1160, 1030, 945; ^{1}H (δ, ppm; D$_2$O): 7.7 (d, 2 H, ArH), 7.35 (d, 2 H, ArH), 5.1 (s, 7 H, CH), 3.9–3.8 (m, 28 H, CH and CH_2), 3.8–3.7 (m, 7 H, CH), 3.7–3.5 (m, 14 H, CH), 2.4 (s, 3 H, CH_3).

2.7 Catenanes and Rotaxanes

Catenanes are fascinating examples of molecular knots in which one cyclic molecule penetrates another, as shown in Figure 2.19. The first example was

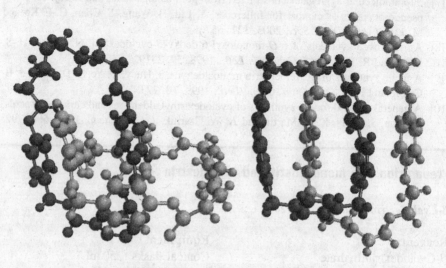

Figure 2.19 Examples of catenanes

reported by Wasserman in 1960 [1]; however, the recent resurgence in interest is due largely to the advances made by the groups of Sauvage and Stoddart, both of which rely on large polyether rings for one or more components. The approach taken by Sauvage [2] was to use tetrahedral copper(I) to initiate the formation of the interlocked rings. Derivatives of 1,10-phenanthroline bind to copper(I) in a 2:1 stoichiometry. The divergent termini of these ligands then react under high dilution conditions with diiodopolyethers to give the catenate in 42 per cent yield. The result is an interlocked pair of macrocycles in which both phenanthrolines remain bound to the copper centre. Treatment with cyanide removes the copper and gives rise to a conformational rearrangement in which the phenanthrolines move as far apart as possible. Sauvage took this trick a step further by introducing a double twist with two 1,10-phenanthroline derivatives in sequence. Once this unwound, following the removal of copper, a molecular trefoil knot was formed [2]. To show that the principle could be used to form even more exotic species, the Lehn group prepared bipyridyl trimers that spontaneously formed double helical complexes in the presence of copper(I) salts [3].

Stoddart's extensive work on catenanes started with the realization that diben-zocrown ethers contained two environments, the polyether electron-donor region and the electron-rich region with π-stacking potential. The aromatic regions were used to hold a 4,4'-bipyridinium-based molecular clip in place while cyclization with 1,4-di(bromomethyl)benzene was effected. The interlocked complex was obtained in 70 per cent yield as the tetracationic salt with hexafluorophosphate counterions [4].

Analysis of NMR data showed that the pyridinium 'molecular shuttle' travels around the polyether thread and has an affinity for the aromatic 'stations' through π–π-stacking between the pyridinium groups in the shuttle and the electron-rich aromatic stations. The shuttle also spins around the thread under a secondary process. The same synthetic strategy has been used to create larger polyethers containing more stations. Reaction of tetra(1,4-benzo)[68]crown-20 with the same molecular clip and 1,4-dibromobenzene under different sets of conditions gives catenanes with one or two shuttles. Perhaps the most spectacular example of this strategy was the synthesis of the [5]catenane, Olympiadane [5].

Related to the catenanes are the rotaxanes, molecules in which one linear molecule is threaded through a macrocycle and held in place by large 'stoppers' at either end that are too large for the macrocyclic 'bead' to fit through, as illustrated in Figure 2.20. Many combinations of macrocycles and threads are now known to form rotaxanes, even the seemingly tight-fitting combination of α-cyclodextrin and 1,1''-(α, ω-alkanediyl)bis(4,4'-bipyridinium) salts [6]. The idea of threading one

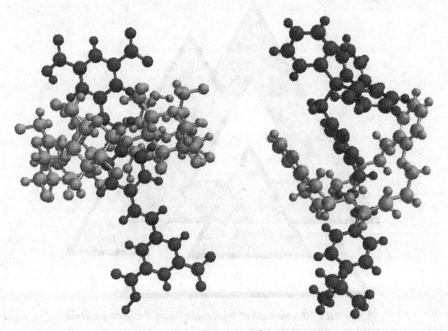

Figure 2.20 Examples of rotaxanes

molecule through another has been around since an early paper by Harrison [7] but it is thanks again to the Stoddart group that this concept has gained popularity over the past 15 years or so. The first full paper in an ever-expanding output on catenanes and rotaxanes was published in 1992 and set the agenda for the concept of 'molecular meccano' [8]. In essence this is a toolkit of functional molecules that can be combined to make macrocycles or molecular threads with different properties linked in a well-defined manner. Thus a macrocycle with electron-deficient aromatic units, often bipyridinium as in the above example, could be formed through a templated reaction around an electron rich aromatic group such as a dihydroxybenzene spacer in a large benzocrown or polyether. The smaller macrocycle is then free to rotate around, and travel along, the larger compound but is held on the molecular thread and cannot be removed without breaking covalent bonds. If the larger molecule contains other functional groups that change electron density, the small cycle can shuttle between complementary sites. The application of this principle to linear molecules that have bulky termini to prevent the shuttle from slipping off the ends has led to a greater understanding of the compounds pioneered by Harrison and led to a greater appreciation of these rotaxane type of molecular entanglements [9,10]. Notable among other groups to have investigated rotaxane-forming systems are those of Leigh [11,12] and Gibson [13], although interest in this class of supramolecules has become more widespread. In addition to crown ether or cycloamide shuttles there is also interest in rotaxanes from

Figure 2.21 An example of Borromean rings: the Viking *valknut*

cyclodextrin chemists [14] to the extent that an entire review dedicated to this subclass of rotaxanes has been published [15].

In an exciting extension of catenane chemistry the groups of Stoddart and Atwood recently reported the synthesis and crystal structure of a triply-interlocked supramolecule that forms a Borromean link. The Borromean topology, known from ancient times through Norse (the Viking *valknut*) and early Christian symbolism, has three interlocked rings, as the design in Figure 2.21 indicates. The three rings interpenetrate each other and cannot be separated unless one ring is broken, in which case all three rings fall apart. The motif has been observed in DNA-knots prepared by Seeman but has proved elusive in synthetic organic

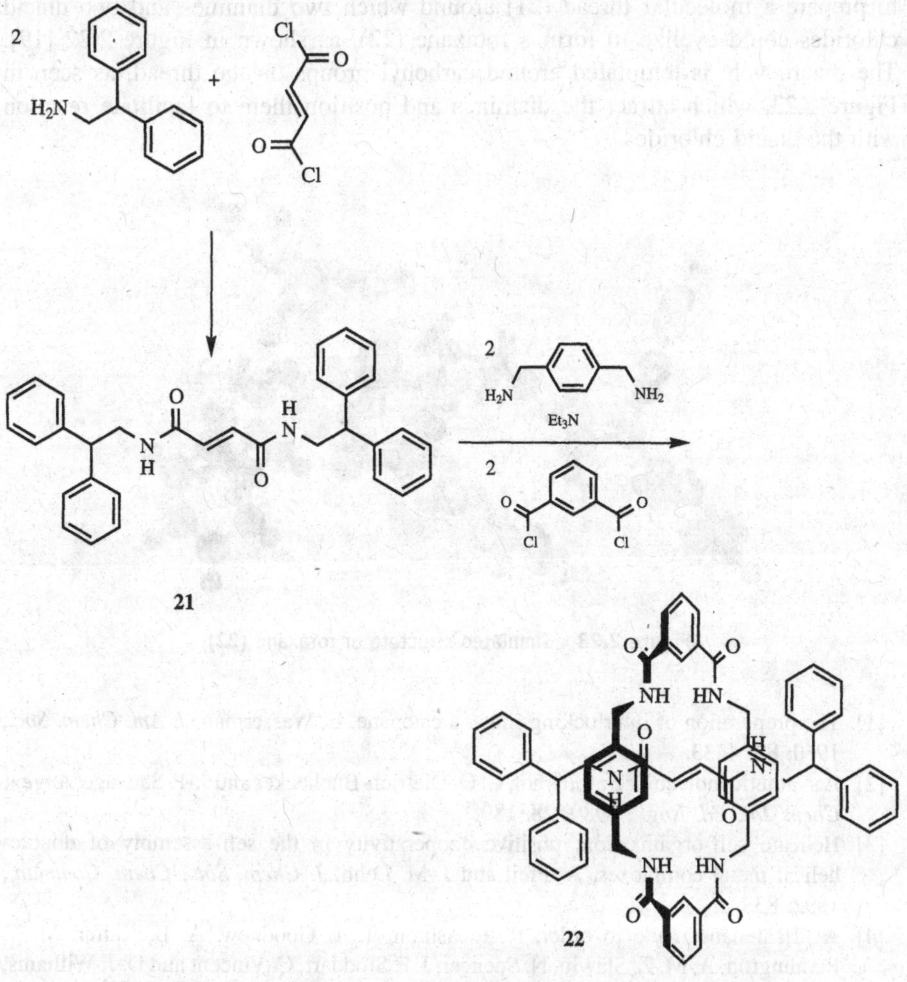

Figure 2.22 Synthesis of the fumaramide thread (**21**) and rotaxane (**22**)

chemistry. In Stoddart's example, each ring is prepared by a zinc-templated cyclization of two 2,4-diformylpyridines and two diamines containing bipyridyl ligating sites. The interpenetrating nature of the rings was confirmed by X-ray crystallography [16].

In a further example of interpenetrating systems, the group of Böhmer reported the synthesis and structure of an [8]catenane that forms when two calix[4]arenes self-assemble into a molecular capsule and are then locked in place through reactions of upper-rim substituents [17].

The example given here comes from the outstandingly productive approach pioneered by Leigh. The group originally prepared a self-templating catenane in a quite astonishing 22 per cent yield from a diaminoaromatic system and a diacid chloride [18]. Further work revealed how the templated method could be extended to prepare a molecular thread (21) around which two diamines and two diacid chlorides could cyclize to form a rotaxane (22), as shown in Figure 2.22 [19]. The macrocycle is templated around carbonyl groups on the thread, as seen in Figure 2.23, which attract the diamines and position them to facilitate reaction with the diacid chlorides.

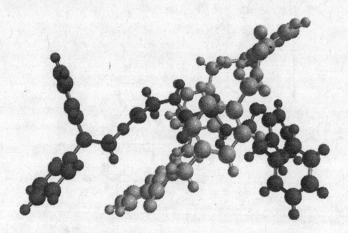

Figure 2.23 Simulated structure of rotaxane (22)

[1] The preparation of interlocking rings: a catenane, E. Wasserman, *J. Am. Chem. Soc.*, 1960, **82**, 4433.

[2] A synthetic molecular trefoil knot, C. O. Dietrich-Buchecker and J.-P. Sauvage, *Angew. Chem. Int. Ed. Engl.*, 1989, **28**, 189.

[3] Helicate self-organisation: positive cooperativity in the self-assembly of double-helical metal complexes, A. Pfeil and J.-M. Lehn, *J. Chem. Soc., Chem. Commun.*, 1992, 838.

[4] A [2]catenane made to order, P. R. Ashton, T. T. Goodnow, A. E. Kaifer, M. V. Reddington, A. M. Z. Slawin, N. Spencer, J. F. Stoddart, C. Vincent and D. J. Williams, *Angew. Chem., Int. Ed. Engl.*, 1989, **28**, 1396.

[5] Olympiadane, D. B. Amabilino, P. R. Ashton, A. S. Reder, N. Spencer and J. F. Stoddart, *Angew. Chem., Int. Ed. Engl.*, 1994, **33**, 1286.

[6] [2]Rotaxane molecular shuttles employing 1,2-bis(pyridinium)ethane binding sites and dibenzo-24-crown-8 ethers, S. J. Loeb and J. A. Wisner, *Chem. Commun.*, 2000, 1939.

[7] The synthesis of a stable complex of a macrocycle and a threaded chain, I. T. Harrison and S. Harrison, *J. Am. Chem. Soc.*, 1967, **89**, 5723.

[8] Molecular Meccano 1. [2]Rotaxanes and a [2]catenane made to order, P. L. Anelli, P. R. Ashton, R. Ballardini, V. Balzani, M. Delgado, M. T. Gandolfi, T. T. Goodnow, A. E. Kaifer, D. Philp, M. Pietraszkiewicz, L. Prodi, M. V. Reddington, A. M. Z. Slawin, N. Spencer, J. F. Stoddart, C. Vicent and D. J. Williams, *J. Am. Chem. Soc.*, 1992, **114**, 193.

[9] *Catenanes, Rotaxanes, and Knots*, G. Schill, Academic Press, New York, 1971.

[10] Interlocked and intertwined structures and superstructures, D. B. Amabilino and J. F. Stoddart, *Chem. Rev.*, 1995, **95**, 2725.

[11] The synthesis and solubilization of amide macrocycles via rotaxane formation, A. G. Johnson, D. A. Leigh, A. Murphy, J. P. Smart and M. D. Deegan, *J. Am. Chem. Soc.*, 1996, **118**, 10662.

[12] Peptide-based molecular shuttles, A. S. Lane, D. A. Leigh and A. Murphy, *J. Am. Chem. Soc.*, 1997, **119**, 11092.

[13] Polyrotaxanes based on polyurethane backbones and crown-ether cyclics. 1. Synthesis, Y. X. Shen, D. H. Xie and H. W. Gibson, *J. Am. Chem. Soc.*, 1994, **116**, 537.

[14] Self-assembling metal rotaxane complexes of α-cyclodextrin, R. S. Wylie and D. H. Macarney, *J. Am. Chem. Soc.*, 1992, **114**, 3136.

[15] Cyclodextrin-based catenanes and rotaxanes, S. A. Nepogodiev and J. F. Stoddart, *Chem. Rev.*, 1998, **98**, 1959.

[16] Molecular Borromean rings, K. S. Chichak, S. J. Cantrill, A. R. Pease, S.-H. Chiu, G. W. V. Cave, J. L. Atwood and J. F. Stoddart, *Science*, 2004, **304**, 1308.

[17] Multiple catenanes derived from calix[4]arenes, L. Wang, M. O. Vysotsky, A. Bogdan, M. Bolte and V. Böhmer, *Science*, 2004, **304**, 1312.

[18] Facile synthesis and solid-state structure of a benzylic amide [2]catenane, A. G. Johnson, D. A. Leigh, R. J. Pritchard and M. D. Deegan, *Angew. Chem. Int. Ed. Engl.*, 1995, **34**, 1209.

[19] Stiff, and sticky in the right places: the dramatic influence of preorganizing guest binding sites on the hydrogen bond-directed assembly of rotaxanes, F. G. Gatti, D. A. Leigh, S. A. Nepogodiev, A. M. Z. Slawin, S. J. Treat and J. K. Y. Wong, *J. Am. Chem. Soc.*, 2001, **123**, 5983.

Preparation of a rotaxane

Fumaryl bisamide thread (21)

Reagents
2,2-Diphenylethylamine
Fumaryl chloride [HARMFUL]
Dry chloroform [TOXIC]

Equipment
Round-bottomed flask (250 mL)
Magnetic stirrer and stirrer bar
Büchner funnel and flask

Note: Work in a fume hood, particularly in view of the large amount of chloroform used.

Dissolve 2,2-diphenylethylamine (1.0 g, 5 mmol) in dry chloroform (60 mL) and add dropwise over 90 min to a stirred solution of fumaryl chloride (0.27 mL, 0.4 g, 2.5 mmol) in dry chloroform (40 mL) in a 250 mL round-bottomed flask. Once the addition is complete stir for a further 30 min then remove solvent under reduced pressure until the solution turns cloudy. Warm the mixture until the mixture clarifies then leave in a freezer for *ca.* 24 h. The resulting precipitate was isolated by filtration to yield the fumaryl bisamide thread (**21**) as white fibrous crystals.

Yield: 0.6 g, (50%); m.p.: >250 °C; IR (v, cm^{-1}): 3265, 3075, 1630, 1555, 1195, 695; ^{1}H $(\delta, \text{ppm}; \text{CDCl}_3)$: 7.4–7.2 (m, 20 H, Ar*H*), 6.7 (s, 2 H, =C*H*R), 5.7–5.6 (br s, 2 H, N*H*), 4.2 (m, 2 H, C*H*Ar$_2$), 4.0 (m, 4 H, C*H*$_2$).

Rotaxane (22)

Reagents	**Equipment**
Fumaryl bisamide (**21**)	Three-necked round-bottomed flask (50 mL)
p-Xylylenediamine [CORROSIVE]	Magnetic stirrer and stirrer bar
Isophthaloyl chloride [CORROSIVE]	Inert atmosphere line
Triethylamine [FLAMMABLE;	Syringes (5 mL) and septa
CORROSIVE]	Addition funnels
Dry chloroform [TOXIC]	Büchner funnel and flask
Dry acetonitrile [FLAMMABLE]	
Dimethylformamide (DMF)	
[HARMFUL]	
Distilled water	

Note: Work in a fume hood wherever possible.

Prepare solutions of *p*-xylylenediamine (0.115 g, 0.84 mmol) in chloroform (5 mL) and isophthaloyl chloride (0.170 g, 0.84 mmol) in chloroform (5 mL). Under an inert atmosphere set up the reaction vessel. Thoroughly dry the round-bottomed flask and fit it with a T-shaped glass fitting that will allow the inert gas to fill, then blow through, the flask. Place rubber septa in the two other necks. Using syringes pushed through the septa, add the solutions dropwise and simultaneously over 30 min to a stirred solution of fumaryl amide (**21**) (0.100 g, 0.21 mmol) and triethylamine (0.22 mL, 0.16 g, 1.6 mmol) in a mixture of acetonitrile (4 mL) and chloroform (9 mL). Note that that the amide may require gentle heating to get it into solution. Once the addition is complete stir the reaction mixture for a further 3 h before filtration. Remove solvents under reduced pressure to leave a white solid. Dissolve the crude product in DMF (*ca.* 3 mL) and add distilled water dropwise (*ca.* 5 mL or until the solution becomes slightly cloudy). Crystals of rotaxane (**22**) are formed overnight and can be isolated by filtration.

Yield: 0.25 g, (75%); m.p.: >250 °C; IR (v, cm^{-1}): 3480, 3305, 1635, 1535, 700; ^1H (δ, ppm; DMSO-d_6): thread 8.15 (m, 2 H, NH), 7.3–7.1 (m, 20 H, ArH), 5.65 (s, 2 H, =CHR), 4.1 (m, 2 H, CHAr$_2$), 3.6 (m, 4 H, CH_2); shuttle 8.65 (s, 2 H, ArH), 8.65–8.55 (m, 4 H, NH), 8.1 (d, 4 H, ArH), 7.7 (t, 2 H, ArH), 6.6 (s, 8 H, ArH), 4.2 (s, 8 H, CH_2).

3

Molecular Baskets, Chalices and Cages

3.1 One for Beginners

The facile condensation reaction between formaldehyde and phenols or their derivatives provides a major route into rigid macrocycles used in supramolecular chemistry. Calixarenes, the best-known class of phenol-derived macrocycles, are prepared this way, as are spherands and their relatives. Cyclotriveratrylene, however, is an excellent exemplar of the 'molecular basket' type of ligand and has been known for the best part of a century. The basic cyclotriveratrylene synthesis is shown in Figure 3.1. The original procedure by Mrs. Gertrude Maud Robinson [1] has since been refined by others and many variations are known [2,3].

Despite the early discovery of cyclotriveratrylene it took until 1965 for its structure to be determined by nuclear magnetic resonance (NMR) [4]. A solvated crystal structure only appeared in 1979 [5] and it was a further decade before the solvent-free structure was reported [6]. Many derivatives have been prepared and some exhibit useful inclusion properties such as that in Figure 3.2. The synthesis included here is based on that reported by Robinson. In her paper the structure of the colourless crystalline precipitate resulting from the condensation of veratrole with formaldehyde in sulphuric acid was given a dimeric, rather than trimeric, configuration. At the time the product was believed to be 2,3,6,7-tetramethoxy-9,10-dihydroxyanthracene; however, as this was determined by the percentage of carbon and hydrogen in the compound it is an understandable error as the composition would be the same for any higher homologue.

The procedure is the result of much trial and error together with careful observation: as will be seen from the literature surrounding acid-catalysed reactions between aldehydes and phenols (or their derivatives such as resorcinol,

A Practical Guide to Supramolecular Chemistry   Peter J. Cragg
© 2005 John Wiley & Sons, Ltd

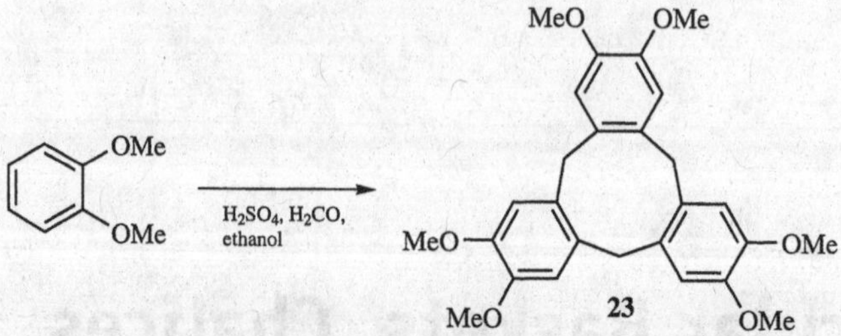

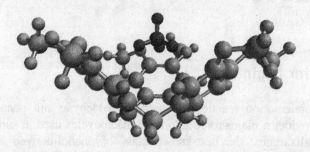

Figure 3.1 Synthesis of cyclotriveratrylene (**23**)

Figure 3.2 Structure of a cyclotriveratrylene inclusion complex

pyrogallol and veratrole), yields are highly variable. Collet's review [2] lists 15 sets of conditions that give between 21 and 89 per cent yield and, of these examples, six are 'simple' condensation reactions of veratrole and formaldehyde. It appears that the bright colours that the reactions initially produce are due to intermediate species and that, under the wrong conditions, they may form intractable polymers. Low temperatures from the outset seem to be important though, once the product has formed, higher temperatures are necessary to separate the product from remaining reactants and side-products. Recrystallization often results in the formation of fine colourless needles.

[1] A reaction of homopiperonyl and of homoveratryl alcohols, G. M. Robinson, *J. Chem. Soc., Trans.*, 1915, **107**, 267.

[2] Cyclotriveratrylenes and cryptophanes, A. Collet, *Tetrahedron*, 1987, **43**, 5725.

[3] Clean, efficient syntheses of cyclotriveratrylene (CTV) and tris-(*o*-allyl)CTV in an ionic liquid, J. L. Scott, D. R. MacFarlane, C. L. Raston and C. M. Teoh, *Green Chem.*, 2000, **2**, 123.

[4] Conformation of cyclotriveratrylene by nuclear magnetic resonance measurements, B. Miller and B. D. Gesner, *Tetrahedron Lett.*, 1965, **6**, 3351.

[5] The crystal structure of the inclusion compound between cycloveratril, benzene and water, S. Cerrini, E. Giglio, F. Mazza and N. V. Pavel, *Acta Cryst. B*, 1979, **35**, 2605.

[6] Crystal and molecular structure of cyclotriveratrylene, H. Zhang and J. L. Atwood, *J. Cryst. Spec. Res.*, 1990, **20**, 465.

Preparation of cyclotriveratrylene

Cyclotriveratrylene (23)

Reagents
Veratrole
Sulphuric acid (70%) [CORROSIVE]
Aqueous formaldehyde (37%) [TOXIC]
Ethanol [FLAMMABLE]
Toluene [FLAMMABLE]

Equipment
Round-bottomed flask (150 mL)
Syringe (20 mL) and septum
Magnetic stirrer and stirrer bar
Heat gun
Ice bath
Glassware for recrystallization

Note: Work wherever possible in a fume hood.

Add veratrole (10 mL, 10.84 g, 78.5 mmol) to ice-cold sulphuric acid (22 mL, 70 per cent aqueous solution) in a 150 mL round-bottomed flask and stir vigorously. The solution should remain colourless. Add aqueous formaldehyde (12.5 mL, 37 per cent aqueous solution) dropwise using a syringe over 1 h, again with vigorous stirring. The syringe needle should be pushed through a septum that has a second needle inserted to release the pressure. It is important that the formaldehyde is added slowly enough to avoid the mixture heating up. If this occurs then dark purple or black solids appear. A slow rate of addition generates a flocculent off-white precipitate with traces of a mauve material that breaks up to leave an off-white solid. Again, it is vital that stirring is vigorous or the formaldehyde will fail to mix properly. By the end of the addition, about 45 min, the reaction mixture becomes a paste. Leave the paste for 2 h then add ethanol (75 mL), stir and warm the mixture briefly with a heat gun to break up any large solid lumps. Upon cooling, filter the mixture, wash with ethanol (3 × 50 mL or until washings are colourless) and then diethyl ether (50 mL). The crude product is obtained in quantitative yield (20 g) as an off-white powder. Recrystallization of the precipitate from toluene (10 mL per g of crude product) yields cyclotriveratrylene (**23**) as a white solid.

Yield: 4.7 g (40%); m.p.: 222–228 °C; IR (v, cm^{-1}): 1515, 1265, 1090; ^{1}H NMR (δ, ppm; CDCl$_3$),: 6.8 (s, 6 H, Ar*H*), 4.8, 3.6 (dd, 6 H, C*H*$_2$), 3.8 (s, 18 H, C*H*$_3$).

3.2 Calixarenes – Essential Supramolecular Synthons

Calixarenes have been synthesized, knowingly or unknowingly, since the 1870s [1], with a specific interest arising in the 1940s [2], yet reproducible syntheses and unequivocal structural assignments have only appeared since the 1980s [3,4]. The supramolecular and biomimetic behaviour of this class of macrocycles also has a long history: a paper by Cornforth from 1955 reports the antitubercular effects of calixarene alkyl ethers [5]. At the time it was thought that the compounds were calix[4]arene derivatives though latter work revealed them to be calix[8]arenes [6]. Indeed, calixarenes, and in particular calix[4]arenes, have become closely identified with supramolecular chemistry ever since Gutsche optimized conditions for their syntheses. Several reviews, monographs and books have been published on the history and uses of the calixarenes, making extensive discussion here unnecessary [7–10]. In brief, the calixarenes are 2,6-metacyclophanes derived from phenols with a single methylene group separating each phenolic moiety. Many groups can be introduced into the 4-position, or upper rim as it is usually known, and calixarenes with four phenolic groups, together with many higher homologues, have been prepared as shown in Figure 3.3. The smaller calix[3]arenes have been reported but never reproduced [11] though the expanded oxa- and azacalix[3]arenes are well known. The phenolic oxygens can be used to append a variety of ethers or esters, known more generally as lower rim substituents. For most calixarenes the complete nomenclature comprises a prefix, denoting the upper rim substituents, followed by calix[n]arene where n is the

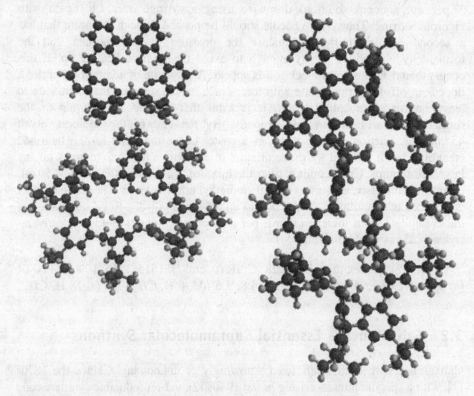

Figure 3.3   Calixarene structures: calix[5]arene (top left), 4-*t*-butylcalix[6]arene (bottom left), 4-*t*-butylcalix[12]arene (right)

number of phenolic units in the cyclic compound. A suffix, describing the lower rim substituents, may also be given. Thus the most commonly encountered member of the class has the designation 4-*t*-butylcalix[4]arene.

The parent calixarenes are flexible during their high temperature synthesis, with rotation of the phenolic moieties about the bridging CH_2 groups possible, but the smallest members of the class 'freeze out' upon cooling to ambient temperatures. This is an important consideration when working with calix[4]arenes as they exist in four conformers that are hard to interconvert and become immobilized in a particular conformer if lower rim substituents are appended, even if reheated to relatively high temperatures. Four principal conformers are observed at room temperature. If all four upper rim substituents are in the same orientation then a *cone* conformer results. If one phenolic group is inverted with respect to the others a *partial cone* conformer is found. Finally, two possibilities exist when two phenol rings are inverted: *1,2 alternate* and *1,3 alternate*. All four conformers are illustrated in Figure 3.4. Similar descriptions exist for larger calixarenes although

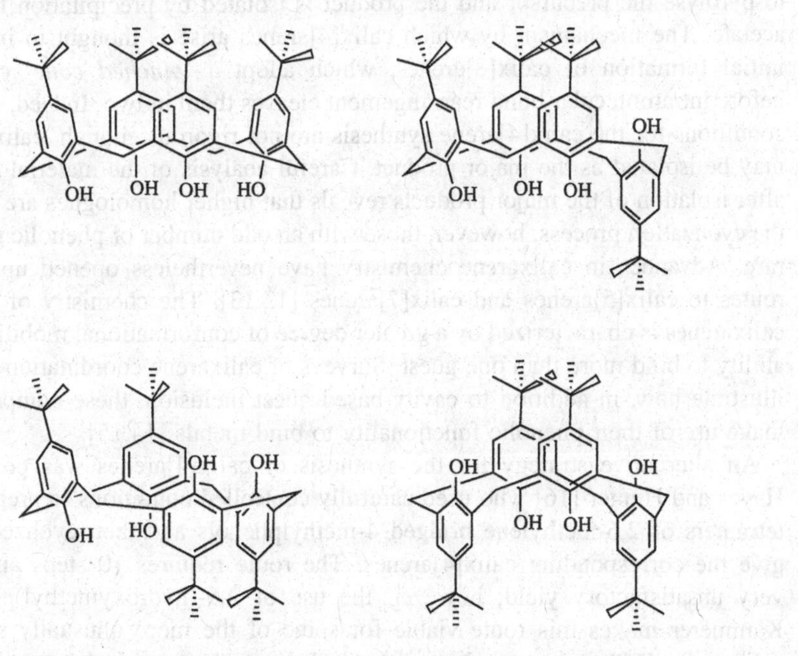

Figure 3.4 Calix[4]arene conformers: *cone* (top left), *partial cone* (top right), *1,2-alternate* (bottom left), *1,3-alternate* (bottom right)

these compounds are often conformationally dynamic and only frozen out in the solid state or in low temperature experiments.

The conventional synthesis of 4-*t*-butylcalix[4]arene is shown in Figure 3.5. This method requires the initial preparation of a precursor by melting 4-*t*-butylphenol in the presence of formaldehyde under basic conditions then dissolving

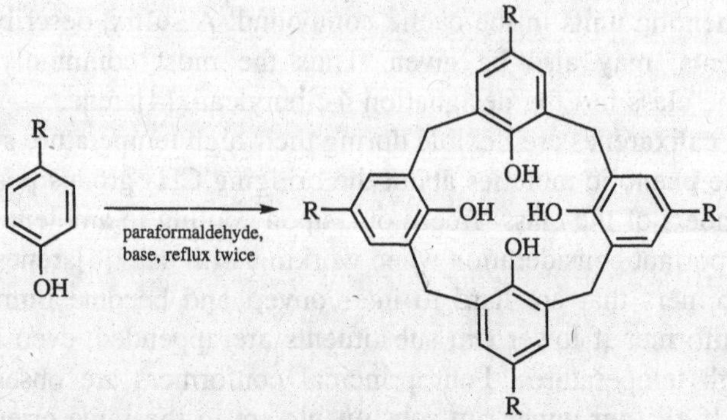

Figure 3.5 Conventional calixarene synthesis

the mixture in diphenyl ether. The reaction mixture is then heated under nitrogen to pyrolyse the precursor and the product is isolated by precipitation from ethyl acetate. The mechanism by which calix[4]arenes arise is thought to involve the initial formation of calix[8]arenes, which adopt a '*pinched cone*' conformer, before intramolecular bond rearrangement cleaves them in two. Indeed, if reaction conditions for the calix[4]arene synthesis are not rigorous enough, calix[8]arenes may be isolated as the major product. Careful analysis of the material remaining after isolation of the major products reveals that higher homologues are formed in the cyclization process, however, those with an odd number of phenolic groups are rare. Advances in calixarene chemistry have nevertheless opened up synthetic routes to calix[5]arenes and calix[7]arenes [12,13]. The chemistry of the larger calixarenes is characterized by a greater degree of conformational mobility and the ability to bind more than one guest. Surveys of calixarene coordination chemistry illustrate how, in addition to cavity-based guest inclusion, these compounds can make use of their phenolic functionality to bind metals [14,15].

An alternative strategy for the synthesis of calix[4]arenes was proposed by Hayes and Hunter [16] who used carefully controlled conditions to prepare linear tetramers of 2,6-methylene bridged 4-methylphenols and then cyclized them to give the corresponding calix[4]arenes. The route requires 10 steps and gives a very unsatisfactory yield; however, the use of 2,6-(hydroxymethyl)phenols by Kämmerer makes this route viable for some of the more unusually substituted calixarenes [17]. A more imaginative synthetic route recently appeared in which the calixarene is prepared from a biscarbene- and a diyne-substituted phenol through a concerted triple benzannulation that generates two further aromatic rings [18]. The advantage of this approach is that it may be used to synthesize a wide range of asymmetric calixarenes in 20 to 40 per cent yield. It is also possible to prepare 4-*t*-butylcalix[4]arene simply from dehydrating 4-*t*-butyl-2,6-(hydroxymethyl)phenol in an aromatic solvent, as shown in Figure 3.6, but only if the precursor has been carefully recrystallized to remove any traces of acid that would otherwise catalyse the formation of oxacalixarenes [19]. This method has yet to be

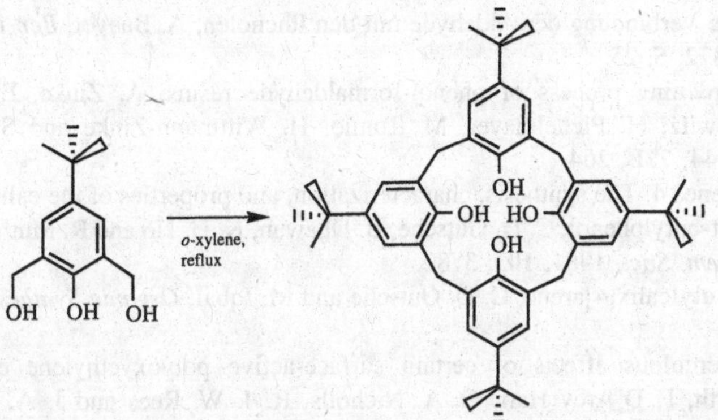

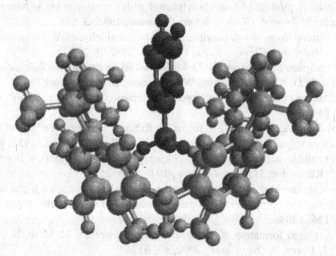

Figure 3.6   Synthesis of 4-*t*-butylcalix[4]arene from bis(hydroxymethyl)-4-*t*-butylphenol

optimized although it may become an attractive alternative to the standard synthesis in the future [4].

No synthesis for this compound is included simply because different-sized calixarenes and many of their derivatives are now readily available from standard suppliers of fine chemicals at very reasonable prices. Given the cost of the diphenyl ether used in their synthesis, it is generally preferable to buy the 4-*t*-butylcalix[n]arenes unless they are required on an industrial scale.

Two attributes of calix[4]arenes have brought them very much to the fore in the supramolecular field: their inclusion properties and potential to form into sheets or spheres in the solid state. Although both of these attributes are found to a great degree in the chemistry of calix[4]arene derivatives, such as 4-sulphonatocalix[4]-arene or 4-*t*-butylcalix[4]arenetetraacetamide, they are largely absent from the parent compound which forms a limited range of inclusion compounds. One of the most commonly encountered, the toluene complex, is illustrated in Figure 3.7.

Figure 3.7   Simulated structure of a 4-*t*-butylcalix[4]arene inclusion complex

[1] Uber die Verbindung der Aldehyde mit den Phenolen, A. Baeyer, *Ber. Dtsch. Chem. Ges.*, 1872, **5**, 25.

[2] The hardening process of phenol-formaldehyde resins, A. Zinke, E. Ziegler, E. Martinowitz, H. Pichelmayer, M. Tomio, H. Wittmann-Zinke and S. Zwanziger, *Ber.*, 1944, **77B**, 264.

[3] Calixarenes 4. The synthesis, characterization, and properties of the calixarenes from para-tert-butylphenol, C. D. Gutsche, B. Dhawan, K. H. No and R. Muthukrishnan, *J. Am. Chem. Soc.*, 1981, **103**, 3782.

[4] *p-tert*-Butylcalix[4]arene, C. D. Gutsche and M. Iqbal, *Organic Syntheses*, 1990, **68**, 234.

[5] Antituberculous effects of certain surface-active polyoxyethylene ethers, J. W. Cornforth, P. D'Arcy Hart, G. A. Nicholls, R. J. W. Rees and J. A. Stock, *Br. J. Pharmacol.*, 1955, **10**, 73.

[6] X-ray crystal and molecular structure of the para-tert-butylphenol-formaldehyde cyclic octamer cyclo(octa[(5-tert-butyl-2-acetoxy-1,3-phenylene)methylene]), G. D. Andreetti, R. Ungaro and A. Pochini A, *J. Chem. Soc., Chem. Commun.*, 1981, 533.

[7] *Calixarenes*, C. D. Gutsche, Royal Society of Chemistry, Cambridge, 1989.

[8] *Calixarenes Revisited*, C. D. Gutsche, Royal Society of Chemistry, Cambridge, 1997.

[9] Calixarenes, C. D. Gutsche, *Acc. Chem. Res.*, 1983, **16**, 161.

[10] Calixarenes, macrocycles with (almost) unlimited possibilities, V. Böhmer, *Angew. Chem. Int. Ed. Engl.*, 1995, **34**, 713.

[11] The synthesis of 5,11,17-trihalotetracyclo[13 3 1 1-3,7 1-9,13]henicosa-1(19),3,5,7 (20),9,11,13(21),15,17-nonaene-19,20,21-triols and 5,11,17-trihalo-19,20,21-trihydroxytetracyclo[13 3 1 1-3,7 1-9,13]henicosa-1(19),3,5,7(20),9,11,13(21),15,17-non-aene-8,14-dione [1] – cyclo-derivatives of phloroglucide analogs, A. A. Moshfegh, E. Beladi, L. Radnia, A. S. Hosseini, S. Tofigh and G. H. Hakimelahi, *Helv. Chim. Acta*, 1982, **65**, 1264.

[12] Formaldehyde polymers 29. Isolation and characterization of calix[5]arene from the condensation product of 4-tert-butylphenol with formaldehyde, A. Ninagawa and H. Matsuda, *Makromol. Chem. – Rapid Commun.*, 1982, **3**, 65.

[13] Calix[7]arene from 4-tert-butylphenol and formaldehyde, Y. Nakamoto and S. I. Ishida, *Makromol. Chem. – Rapid Commun.*, 1982, **3**, 705.

[14] Metal complexes of calixarenes, D. M. Roundhill, in *Progress in Inorganic Chemistry*, Vol. 43, K. D. Karlin (ed.), John Wiley & Sons Inc., New York, 1995.

[15] Coordination chemistry of the larger calixarenes, C. Redshaw, *Coord. Chem. Rev.*, 2003, **244**, 45.

[16] Phenol-formaldehyde and allied resins VI: Rational synthesis of a 'cyclic' tetranuclear p-cresol novolak, B. T. Hayes and R. F. Hunter, *J. Appl. Chem.*, 1958, **8**, 743.

[17] A new synthetic access to cyclic oligonuclear phenolic compounds, V. Böhmer, P. Chim and H. Kämmerer, *Makromol. Chem.*, 1979, **180**, 2503.

[18] A new convenient strategy for the synthesis of calixarenes via a triple annulation of Fischer carbene complexes, V. Gopalsamuthiram and W. D. Wulff, *J. Am. Chem. Soc.*, 2004, **126**, 13936.

[19] Acid-catalyzed formation of hexahomooxacalix[3]arenes, M. Miah, N. N. Romanov and P. J. Cragg, *J. Org. Chem.*, 2002, **67**, 3124.

3.3 Adding Lower Rim Functionality to the Calixarenes

As it stands, 4-*t*-butylcalix[4]arene does not look like a particularly inspiring supramolecular synthon. It has a well-defined cavity but contains bulky *t*-butyl groups on the upper rim and limited potential to functionalize further through the lower rim phenolic groups. The compound is also insoluble in water, a major drawback if it is to be used in any biological context. Fortunately it is possible to modify both upper and lower rims in order to introduce useful properties. In order to make use of the macrocyclic cavity it is important that following any functionalization the compound retains its cone conformer. The high energy barriers to conformer interconversion ensure that, if the cone conformer of the parent calixarene is employed, a cone derivative will form. This is not the case for the larger homologues in the calix[*n*]arene series nor is it true for the expanded oxa- and azacalix[*n*]arenes. Simple ethers may be prepared easily and are particularly useful if the phenol needs protection. To be more useful the derivatives must incorporate further functionality such as esters or amides. Both can be formed readily to give calixarenes with extended lower rims. Tetraester derivatives of calix[4]arenes were first reported in 1986 [1] to be followed shortly by the amide analogues [2]. The tetraamides are particularly valuable as they form complexes with transition and *p*-block metals, such as the copper complex illustrated in Figure 3.8, in addition to the more usual alkali metals [3]. Many

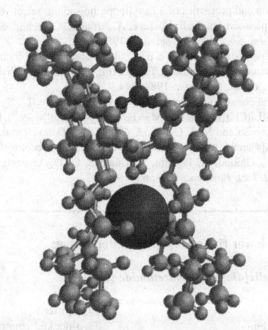

Figure 3.8 Simulated structure of the cone-4-*t*-butylcalix[4]arenetetra(*N,N*-diethylacetamide) copper complex with included acetonitrile

other lower rim derivatives are now common including the calixcrowns that incorporate a bridging polyether between phenol groups.

The preparation of cone-t-butylcalix[4]arenetetra(N,N-diethylacetamide), **25**, using the approach of Ungaro and Ugozzoli [4] as shown in Figure 3.9, is described below. It is a widely versatile calixarene derivative as indicated above and probably has many more possibilities to form inclusion complexes than those listed.

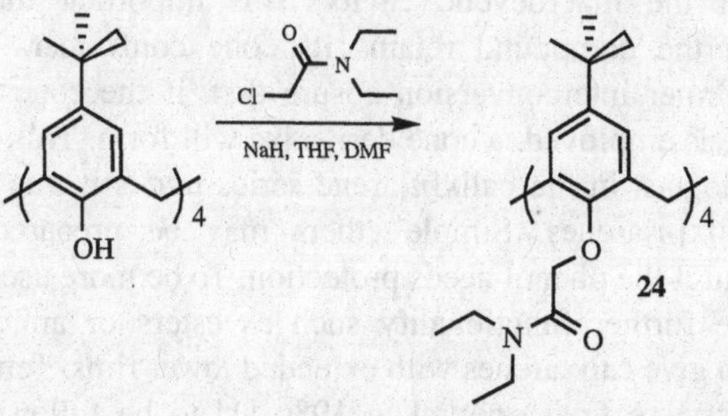

Figure 3.9 Synthesis of cone-t-butylcalix[4]arenetetra(N,N-diethylacetamide) (**24**)

[1] The preparation and properties of a new lipophilic sodium selective ether ester ligand derived from para-tert-butylcalix(4)arene, A. Arduini, A. Pochini, S. Reverberi and R. Ungaro, *J. Chem. Soc., Chem. Commun.*, 1984, 981.

[2] Encapsulated potassium cation in a new calix[4]arene neutral ligand – synthesis and X-ray crystal structure, G. Calestani, F. Ugozzoli, A. Arduini, E. Ghidini and R. Ungaro, *J. Chem. Soc., Chem. Commun.*, 1987, 344.

[3] Versatile cation complexation by a calix[4]arene tetraamide (L). Synthesis and crystal structure of [ML][ClO$_4$]$_2 \cdot n$MeCN (M = Fe^{II}, Ni^{II}, Cu^{II}, Zn^{II} or Pb^{II}), P. D. Beer, M. G. B. Drew, P. B. Leeson and M. I. Ogden, *J. Chem. Soc., Dalton Trans.*, 1995, 1273.

[4] p-t-Butylcalix[4]arene tetra-acetamide: a new strong receptor for alkali cations, A. Arduini, E. Ghidini, A. Pochini, R. Ungarro, G. D. Andreeti, G. Calestani and F. Ugozzoli, *J. Incl. Phenom.*, 1988, **6**, 119.

Preparation of lower rim calix[4]arene derivatives

Cone-4-t-butylcalix[4]arenetetraacetamide (24)

Reagents	**Equipment**
t-Butylcalix[4]arene	Two-necked round-bottomed flask
Tetrahydrofuran (THF) [FLAMMABLE]	(250 mL)

Dimethylformamide (DMF)
Sodium hydride (60% dispersion in
mineral oil) [CORROSIVE; REACTS
VIOLENTLY WITH WATER]
N,N-Diethylchloroacetamide
[TOXIC; LACHRYMATOR]
Distilled water
Hydrochloric acid (1 M) [CORROSIVE]
Dichloromethane [TOXIC]
Diethyl ether [FLAMMABLE]
Hexane [FLAMMABLE; NEUROTOXIN]
Magnesium sulphate

Reflux condenser
Heater/stirrer and stirrer bar
Glass syringe
Inert atmosphere line
Rotary evaporator
Glassware for work-up

Note: Work in a fume hood whenever possible and take appropriate care using sodium hydride.

Suspend 4-*t*-butylcalix[4]arene (1.95 g, 3.0 mmol) in a mixture of dry THF* (80 mL) and dry DMF (20 mL) in a two-necked 250 mL round-bottomed flask under an inert atmosphere. Warm the mixture and stir to dissolve the solid as much as possible then cool to room temperature before adding sodium hydride (1.15 g [60 per cent dispersion in mineral oil], 30 mmol) whereupon the mixture clarifies briefly. Once the effervescence subsides a white precipitate forms, indicating the formation of the sodium salt. Add *N,N*-diethylchloroacetamide (3.4 mL, 3.75 g, 25 mmol) by syringe and reflux the mixture for 4 h using a heating mantle. After cooling to room temperature transfer to a single-necked round-bottomed flask and remove the volatile solvents under reduced pressure. Add water (40 mL) to the alkaline viscous residue and acidify the mixture to pH 1 with 1 M hydrochloric acid (*ca.* 30 mL). Some solids may form but will dissolve during the extraction. Wash with hexane (10 mL) to remove mineral oil used to stabilize the sodium hydride. Extract the product into dichloromethane (50 mL then 3×25 mL) and wash with water (30 mL). Dry the organic phase over magnesium sulphate (*ca.* 1 g) and isolate the crude product by removing the organic solvent under reduced pressure. Addition of diethyl ether (100 mL) to the residue yields 4-*t*-butylcalix[4]arenetetraacetamide (**24**) as colourless microcrystals. Subsequent crops of the product can be obtained from the mother liquor to increase the yield.

Yield: 3.3 g (quantitative); m.p.: 228 °C; IR (υ, cm^{-1}): 3385, 1660, 1200, 1055; ^{1}H NMR (δ, ppm; CDCl$_3$): 6.8 (s, 8 H, Ar*H*), 4.55 (s, 8 H, C*H*$_2$CO), 4.45 (d, 4 H, ArC*H*$_2$), 3.55–3.35 (m, 16 H, NC*H*$_2$), 3.3 (d, 4 H, ArC*H*$_2$), 1.2–1.0 (m, 24 H, NCH$_2$*H*$_3$), 1.1 (s, 36 H, CC*H*$_3$).

*See compound **1** for drying method.

3.4   Adding Upper Rim Functionality to the Calixarenes

The only way to functionalize the upper rim of 4-*t*-butylcalix[4]arene is to remove the *t*-butyl group and replace it with something more useful. Two methods have been reported by which this de-*tert*-butylation, sometimes referred to by the more evocative term of 'neutering', may be achieved. The first method is probably the more conventional of the two and involves the use of aluminium trichloride and phenol. It is in essence a Friedel–Crafts reaction where a *t*-butyl group is attached to either benzene or phenol; the innovation is that the source of the *t*-butyl substituent comes from the calixarene [1]. This 'retro-Friedel–Crafts' reaction is thus an effective method for transalkylation that removes an unwanted alkyl group from calix[4]arenes as shown in Figures 3.10 and 3.11. Moreover, it is possible to subtly alter the reaction conditions and partially de-*tert*-butylate

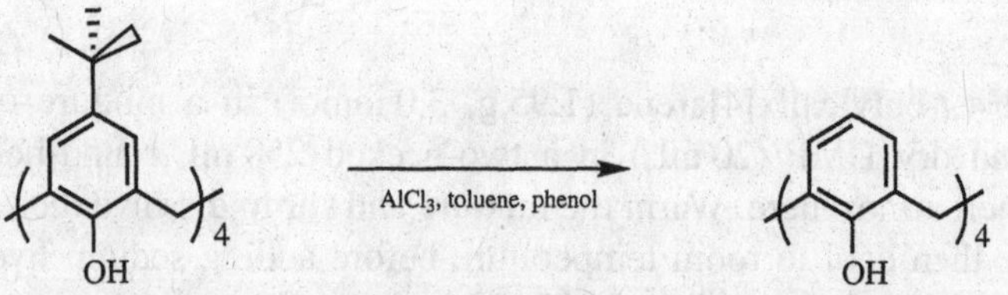

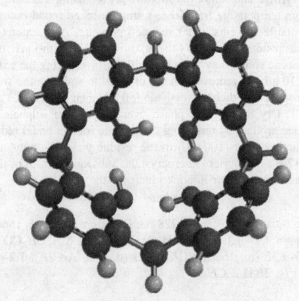

Figure 3.10   De-*tert*-butylation of 4-*t*-butylcalix[4]arene

Figure 3.11   Simulated structure of calix[4]arene (**25**)

dialkoxy-*t*-butylcalix[4]arenes to give other derivatives [2]. The reaction apparently occurs over 18 h in benzene or in 3 h in toluene with added phenol. The method described here avoids using benzene yet affords a relatively straightforward workup and isolation of the calix[4]arene. The alternative method for removing the *t*-butyl groups involves the 'superacid', Nafion™ and also allows for partial dealkylation [3]. Although the reaction is slightly cleaner, it does not seem to have found widespread appeal.

Once the *t*-butyl group has been removed the question of what to put in its place arises. Calix[4]arene can be functionalized in several ways, all related to the well-known chemistry of phenols. It is possible to protect the lower rim phenolic oxygens and brominate [4], as a prelude to further functionalization such as peptide attachment [5], or treat with nitric acid to prepare 4-nitro derivatives [6]. Without any further modification to protect lower rim phenolic positions, one outstanding possibility emerges: sulphonation to the 4-sulphonatocalix[4]arene derivative [7]. This compound, illustrated in Figures 3.12 and 3.13, has the advantages of water solubility and the ability to bind guests through upper and lower oxygens. The synthesis is based on the previously published route to 4-sulphonatocalix[6]arenes [8]. In addition to the extensive structural studies of the 4-sulphonatocalix[4]arenes undertaken by the Atwood group since the 1980s [9–11], there have been kinetic studies [12], complexes formed with amino acids

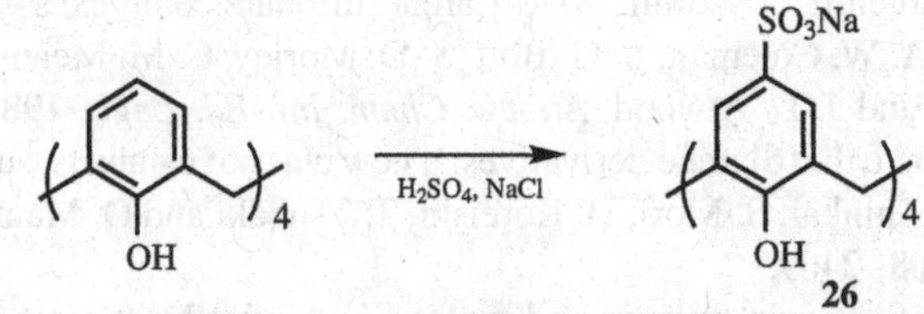

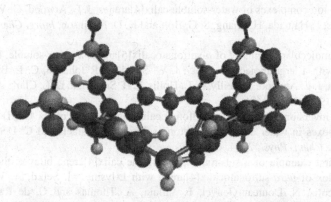

Figure 3.12   Synthesis of 4-sulphonatocalix[4]arene (**26**)

Figure 3.13   Simulated structure of 4-sulphonatocalix[4]arene

[13–15] and even computational studies [16]. 4-Phenylcalixarenes can also be derivatized in a similar manner with sulphuric acid or chlorosulphonic acid [17] to generate water-soluble calixarenes with even deeper cavities.

Given the continued interest in 4-sulphonatocalix[4]arene its synthesis is given here. Both steps, preparation of calix[4]arene (**25**) using the retro-Friedel–Crafts route and subsequent sulphonation to give 4-sulfonatocalix[4]arene (**26**), are high yielding and give highly pure products.

[1] Calixarenes 17. Functionalized calixarenes – the Claisen rearrangement route, C. D. Gutsche, J. A. Levine and P. K. Sujeeth, *J. Org. Chem.*, 1985, **50**, 5802.

[2] Selective functionalization of calix[4]arenes at the upper rim, J.-D. van Loon, A. Arduini, L. Coppi, W. Verboom, A. Porchini, R. Ungaro, S. Harkema and D. N. Reinhoudt, *J. Org. Chem.*, 1990, **55**, 5639.

[3] De-*tert*-butylation of *p-tert*-butylcalix[4]arene with Nafion: a new route to the synthesis of completely and partially debutylated *p-tert*-butylcalix[4]arenes, S. G. Rha and S.-K. Chang, *J. Org. Chem.*, 1998, **63**, 2357.

[4] Calixarenes 6. Synthesis of a functionalizable calix[4]arene in a conformationally rigid cone conformation, C. D. Gutsche and J. A. Levine. *J. Am. Chem. Soc.* 1982, **104**, 2652.

[5] Synthesis of calix[4]arene library substituted with peptides at the upper rim, H. Hioki, Y. Ohnishi, M. Kubo, E. Nashimoto, Y. Kinoshita, M. Samejima and M. Kodama, *Tetrahedron Lett.*, 2004, **45**, 561.

[6] Ipso nitration of *para-tert*-butylcalix[4]arenes, W. Verboom, A. Durie, R. J. M. Egberink, Z. Asfari and D. N. Reinhoudt, *J. Org. Chem.*, 1992, **57**, 1313.

[7] Novel layer structure of sodium calix[4]arenesulfonate complexes – a class of organic clay mimics, A. W. Coleman, S. G. Bott, S. D. Morley, C. M. Means, K. D. Robinson, H. M. Zhang and J. L. Atwood, *Angew. Chem. Int. Ed. Engl.*, 1988, **27**, 1361.

[8] Hexasulfonated calix[6]arene derivatives: a new class of catalysts, surfactants and host molecules, S. Shinkai, S. Mori, H. Koreishi, T. Tsubaki and O. Manabe, *J. Am. Chem. Soc.*, 1986, **108**, 2409.

[9] Intercalation of cationic, anionic and molecular species by organic hosts. Preparation and crystal structure of $[NH_4]_6[calix[4]arenesulfonate][MeOSO_3] \cdot (H_2O)_2$, S. G. Bott, A. W. Coleman and J. L. Atwood, *J. Am., Chem. Soc.*, 1988, **110**, 610.

[10] Metal ion complexes of water-soluble calix[4]arenes, J. L. Atwood, G. W. Orr, N. C. Means, F. Hamada, H. Zhang, S. G. Bott and K. D. Robinson, *Inorg. Chem.*, 1992, **31**, 603.

[11] Supramolecular chemistry of *p*-sulfonatocalix[5]arene: a water-soluble, bowl-shaped host with a large molecular cavity, J. W. Steed, C. P. Johnson, C. L. Barnes, R. K. Juneja, J. L. Atwood, S. Reilly, R. L. Hollis, P. H. Smith and D. L. Clark, *J. Am. Chem. Soc.*, 1995, **117**, 11426.

[12] Ring inversion kinetics of *p*-sulfonatocalix[4]arene and of its Ca(II) and La(III) complexes in water and water–acetone solutions, Y. Israëli and C. Detellier, *Phys. Chem. Chem. Phys.*, 2004, **6**, 1253.

[13] The first example of a substrate spanning the calix[4]arene bilayer: the solid state complex of *para*-sulfonatocalix[4]arene with L-lysine, M. Selkti, A. W. Coleman, I. Nicolis, N. Douteau-Geuvel, F. Villain, A. Thomas and C. de Rango, *Chem. Commun.*, 2000, 161.

[14] A new packing motif for *para*-sulfonatocalix[4]arene: the solid state structure of the *para*-sulfonatocalix[4]arene D-arginine complex, A. Lazar, E. Da Silva, A. Navaza, C. Barbey and A. W. Coleman, *Chem. Commun.*, 2004, 2162.

[15] Confinement of (S)-serine in tetra-*p*-sulfonatocalix[4]arene bilayers, P. J. Nichols and C. L. Raston, *Dalton Trans.*, 2003, 2923.

[16] Molecular dynamics simulations of *p*-sulfonatocalix[4]arene complexes with inorganic and organic cations in water: A structural and thermodynamic study, A. Mendes, C. Bonal, N. Morel-Desrosiers, J. P. Morel and P. Malfreyt, *J. Phys. Chem. B*, 2002, **106**, 4516.

[17] Direct synthesis of calixarenes with extended arms: *p*-phenylcalix[4,5,6,8]arenes and their water-soluble sulfonated derivatives, M. Makha and C. L. Raston, *Tetrahedron Lett.*, 2001, **42**, 6215.

Preparation of upper rim calix[4]arene derivatives

Calix[4]arene (25)

Reagents
t-Butylcalix[4]arene
Phenol [CORROSIVE]
Aluminium trichloride [CORROSIVE; REACTS VIOLENTLY WITH WATER]
Toluene [FLAMMABLE]
Hydrochloric acid (1 M) [CORROSIVE]
Methanol [FLAMMABLE]
Distilled water
Dichloromethane [TOXIC]

Equipment
Two-necked round-bottomed flask (125 mL)
Calcium chloride guard tube
Inert atmosphere line
Rotary evaporator
Glassware for work-up

Note: Work in a fume hood as an unmistakable odour of benzene appears following treatment with aluminium trichloride.

Suspend 4-*t*-butylcalix[4]arene (1.5 g, 2.0 mmol) and phenol (0.28 g, 3.0 mmol) in toluene (30 mL) in a two-necked round-bottomed flask (125 mL) with an inlet providing a flux of inert gas and an outlet fitted with a calcium chloride guard tube. Add anhydrous aluminium trichloride (1.5 g, 11 mmol), which clarifies the suspension slightly, and stir for 4 h. During this time the mixture turns from colourless to yellow then orange. A red oil may be observed to separate. After 4 h, add hydrochloric acid (65 mL of a 1 M aqueous solution) and continue to stir vigorously for a further 1 h. Ensure that all the sticky red oil is removed from the sides of the vessel and is stirred into the biphasic solution. It may be necessary to use a spatula to free this material. The upper aromatic layer will turn yellow and solids will precipitate out within it. Once all the oil has been stirred to yield a powdery product, leave the mixture to settle for 20 min. Separate the upper organic phase, including any white solids, from the aqueous phase then rinse the reaction vessel with dichloromethane (50 mL). Add the dichloromethane wash to the aqueous phase and extract. Repeat the extraction with more dichloromethane (50 mL) and combine the aromatic and dichloromethane extracts. The second addition of dichloromethane dissolves the solids. Dry the organic phase with sodium sulphate, filter and remove the solvent under reduced pressure until a crystalline precipitate starts to form. Cool to room temperature, filter and wash with methanol (*ca.* 5 mL) to obtain calix[4]arene (**25**) as a white powder.

Yield: 1.1 g (70%); m.p.: >250 °C; IR (v, cm^{-1}): 3150, 1595, 1415, 1265, 1195, 775, 755, 735; ^{1}H NMR (δ, ppm; CDCl$_3$): 10.2 (s, 4 H, ArO*H*), 7.1 (m, 8 H, Ar*H*), 6.7 (m, 4 H, Ar*H*), 4.3 (br s, 4 H, ArC*H*$_2$), 3.6 (br s, 4 H, ArC*H*$_2$).

Sodium sulphonatocalix[4]arene (26)

Reagents
Calix[4]arene (**25**)
Concentrated sulphuric acid [CORROSIVE]
Aqueous sodium chloride
Distilled water

Equipment
Round-bottomed flask (25 mL)
Heating/stirring mantle and
  stirrer bar
Air condenser
Glassware for filtration

Note: Work in a fume hood or well-ventilated laboratory.

Heat calix[4]arene, **25**, (0.85 g, 2 mol) in concentrated sulphuric acid (10 mL) at 70 °C in a round-bottomed flask (25 mL) fitted with an air condenser for 4 h by which time a majority of the solids will have dissolved. Carefully filter the reaction mixture through a glass frit into a flask containing saturated aqueous sodium chloride (20 mL). Caution is required as the mixture effervesces and heat is evolved upon mixing. Heat the solution to boiling and then cool to room temperature. Isolate the crystalline product by filtration and redissolve in a minimum amount of hot distilled water (*ca.* 10 mL). Leave to crystallize overnight at *ca.* 4 °C then filter to give the tetrasodium salt of sulfonatocalix[4]arene (**26**) as colourless needles.

Yield: 1.0 g (80%); m.p.: >250 °C; IR (v, cm^{-1}): 3390, 1635, 1160, 1035, 895, 725; ^{1}H NMR (δ, ppm; D$_2$O): 7.6 (s, 8 H, Ar*H*), 4.8 (br s, 8 H, C*H*$_2$).

3.5 Oxacalix[3]arenes

Homooxacalix[n]arenes ('oxacalixarenes') are members of the calixarene family of macrocycles that incorporate one or more ethereal bridges (CH$_2$OCH$_2$) in place of the more usual methylene groups which link the phenol moieties through the 2- and 6-positions.

Oxacalix[3]arenes were first reported in 1962 [1] but it took a further two decades before a reliable synthetic method appeared [2]. Further routes and refinements continue to make these compounds more accessible [3,4]. In the early syntheses the compounds were prepared from 2,6-bis(hydroxymethyl)phenols which can be easily formed by the action of formaldehyde on the appropriate 4-substituted phenol, as shown in Figure 3.14. The most straightforward synthesis involves the acid-catalysed condensation of 2,6-bis(hydroxymethyl)-4-*t*-butylphenol in refluxing *o*-xylene as shown in Figure 3.15. Mixtures of hexahomotrioxacalix[3]arenes and octahomotetraoxacalix[4]arenes are usually formed, however, if *p*-toluenesulphonic acid or acetic acid is used as the catalyst, the hexahomotrioxacalix[3]arene is the sole product with a broad, dish-like shape as shown in Figure 3.16 [3].

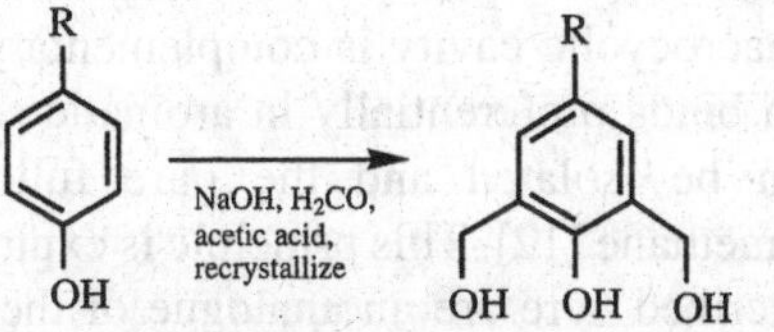

Figure 3.14 Synthesis of 2,6-bis(hydroxymethyl)phenol derivatives (**27**: R = *t*-butyl, **29**: R = phenyl, **33**: R = methyl)

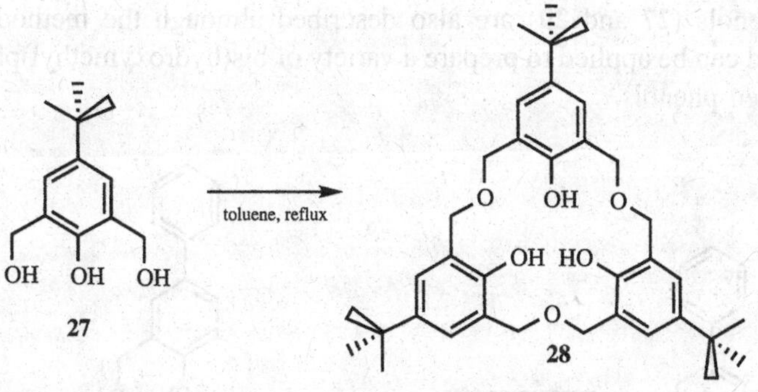

Figure 3.15 Synthesis of 4-*t*-butyloxacalix[3]arene (**28**)

The parent oxacalix[3]arenes show little ability to bind alkali metals, however, a range of quaternary ammonium cations are attracted to the symmetric cavity [4]. Deprotonation of the phenol moieties allows them to bind to transition metals (scandium, titanium, vanadium, rhodium, molybdenum, gold etc.) [5–7], lanthanides (lutetium, yttrium and lanthanum) [8,9] and actinides (uranium, as uranyl)

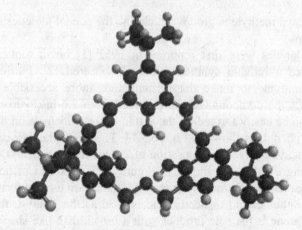

Figure 3.16 Structure of 4-*t*-butyloxacalix[3]arene

[10,11]. One major use has been to purify crude samples of fullerenes. The pseudo-C_3 symmetry of the macrocyclic cavity is complementary to threefold symmetry elements of C_{60} which binds preferentially in aromatic solvents. The oxacalixarene–C_{60} complex can be isolated and the pure fullerene liberated through dissolution in dichloromethane [12]. This principle is explored further in Chapter 5.

The compounds described here are an analogue of the 4-*t*-butylcalix[n]arenes, 4-*t*-butyloxacalix[3]arene (**28**), and a variant, 4-phenyloxacalix[3]arene (**30**, Figures 3.17 and 3.18) containing a deep aromatic cavity. 4-*t*-Butyloxacalix[3]-arene is best prepared by Gutsche's original method [2] despite the more recent publication of several other routes. Syntheses of the respective bis(hydroxy-methyl)phenols (**27** and **29**) are also described although the methods are quite general and can be applied to prepare a variety of bis(hydroxymethyl)phenols from 4-substituted phenols.

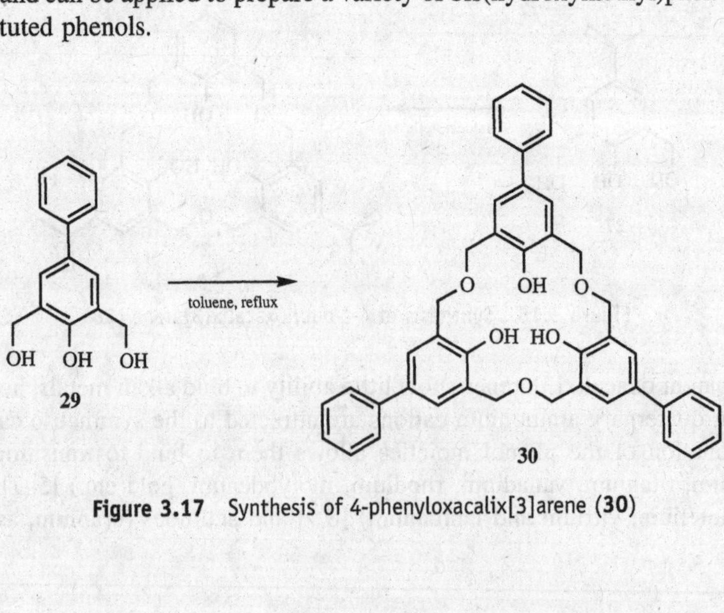

toluene, reflux

29

30

Figure 3.17 Synthesis of 4-phenyloxacalix[3]arene (**30**)

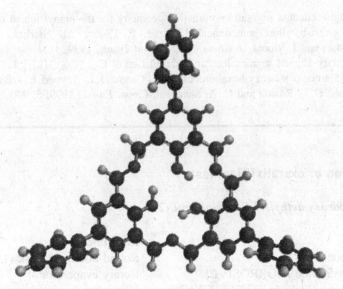

Figure 3.18   Simulated structure of 4-phenyloxacalix[3]arene

[1]   Ring-Kondensate in Alkylphenolharzen, K. Hultzsch, *Kunststoffe*, 1962, **52**, 19.

[2]   Calixarenes 10. Oxacalixarenes, D. Dhawan and C. D. Gutsche, *J. Org. Chem.*, 1983, **48**, 1536.

[3]   A new synthesis of oxacalix[3]arene macrocycles and alkali-metal-binding studies, P. D. Hampton, Z. Bencze, W. D. Tong and C. E. Daitch, *J. Org. Chem.*, 1994, **59**, 4838.

[4]   Acid-catalyzed formation of hexahomooxacalix[3]arenes, M. Miah, N. N. Romanov and P. J. Cragg, *J. Org. Chem.*, 2002, **67**, 3124.

[5]   Linear chain formation by an oxovanadium(V) complex of *p*-methylhexahomotriox-acalix[3]arene, P. D. Hampton, C. E. Daitch, T. M. Alam and E. A. Pruss, *Inorg. Chem.*, 1997, **36**, 2879.

[6]   Titanium complexes of oxacalix[3]arenes – synthesis and mechanistic studies of their dynamic isomerization, P. D. Hampton, C. E. Daitch, T. M. Alam, Z. Bencze and M. Rosay, *Inorg. Chem.*, 1994, **33**, 4750.

[7]   Hexahomotrioxacalix[3]arene: a scaffold for a C_3-symmetric phosphine ligand that traps a hydrido-rhodium fragment inside a molecular funnel, C. B. Dieleman, D. Matt, I. Neda, R. Schmutzler, A. Harriman and R. Yaftian, *Chem. Commun.*, 1999, 1911.

[8]   Selective binding of trivalent metals by hexahomotrioxacalix[3]arene macrocycles: Determination of metal-binding constants and metal transport studies, P. D. Hampton, C. E. Daitch and A. M. Shachter, *Inorg. Chem.*, 1997, **36**, 2956.

[9]   Selective binding of group IIIA and lanthanide metals by hexahomotrioxacalix[3]arene macrocycles, C. E. Daitch, P. D. Hampton, E. N. Duesler and T. M. Alam, *J. Am. Chem. Soc.*, 1996, **118**, 7769.

[10] Trigonal versus tetragonal or pentagonal coordination of the uranyl ion by hexaho-motrioxacalix[3]arenes: solid state and solution investigations, B. Masci, M. Nierlich and P. Thuéry, *New J. Chem.*, 2002, **26**, 120.

[11] An unprecedented trigonal coordination geometry for the uranyl ion in its complex
       with *p-tert*-butylhexahomooxacalix[3]arene, P. Thuéry, M. Nierlich, B. Masci,
       Z. Asfari and J. Vicens, *J. Chem. Soc., Dalton Trans.*, 1999, 3151.
[12] Symmetry-aligned supramolecular encapsulation of C$_{60}$: [C$_{60}$ ⊂ (L)$_2$], L = *p*-benzyl-
       calix[5]arene or *p*-benzylhexahomooxacalix[3]arene, J. L. Atwood, L. J. Barbour, P. J.
       Nichols, C. L. Raston and C. A. Sandoval, *Chem. Eur. J.*, 1999, **5**, 990.

Preparation of oxacalix[3]arenes

2,6-Bis(hydroxymethyl)-4-t-butylphenol (27)

Reagents

p-t-Butylphenol
Sodium hydroxide [CORROSIVE]
Aqueous formaldehyde (37%) [TOXIC]
Tetrahydrofuran (THF) [FLAMMABLE]
Distilled water
Propan-2-ol [FLAMMABLE]
Acetone [FLAMMABLE]
Glacial acetic acid [CORROSIVE]
Toluene [FLAMMABLE]

Equipment

Round-bottomed flask (1 L)
Rotary evaporator
Glassware for filtration and
   recrystallization

Note: Work wherever possible in a fume hood.

Dissolve *p-t*-butylphenol (37.5 g, 25 mmol) in THF (50 mL) in a 1 L round
bottomed flask. Add sodium hydroxide (10 g, 25 mmol, in 30 mL of water) and,
after cooling the mixture, follow this with 37 per cent formaldehyde (45 mL,
55 mmol, 10 per cent excess). Leave the mixture for a week at room temperature
whereupon the solution, initially colourless, will turn orange. At this stage most of
the remaining THF can be removed under reduced pressure to give a darker, more
viscous oil. Add sufficient propan-2-ol (*ca.* 300 mL) to the oil to precipitated out
the sodium 2,6-bis(hydroxymethyl)-4-*t*-butylphenolate salt (30 to 35 g). Filter off
and weigh the precipitate then suspended it in acetone (150 mL). Stir vigorously
and acidify the pale cream salt with a stoichiometric amount of glacial acetic acid
(0.25 mL per g of the sodium salt) dissolved in acetone (10 mL per mL acetic
acid). This precipitates brilliant white sodium acetate, which is easily removed by
filtration. Wash the precipitate with acetone (20 mL) and remove the remaining
solvent to give crude 2,6-bis(hydroxymethyl)-4-*t*-butylphenol (**27**) as an off-white
solid (*ca.* 30 g, 60 per cent). Occasionally the product remains as a pale yellow oil
in which case it can be placed in a freezer overnight to crystallize out. Recrystallization, if required, can be effected by dissolving the solid in refluxing toluene

(1 mL per g) to yield the acid-free product as colourless needle-like crystals. Retaining the mother liquor often yields a second crop.

Yield: 19 g (35%, recrystallized); m.p.: 70 °C; IR (v, cm^{-1}): 3388, 3304, 2950, 2911, 2876, 1483, 1464, 1246, 1060, 1004; ^{1}H NMR (δ, ppm; CDCl$_3$): 7.10 (s, 2 H, Ar*H*), 4.80 (s, 4 H, C*H$_2$*), 1.30 (s, 9 H, CC*H$_3$*).

4-t-Butyloxacalix[3]arene (28)

Reagents
2,6-Bis(hydroxymethyl)-4-*t*-butylphenol (**27**)
o-Xylene [FLAMMABLE]
Dichloromethane [TOXIC]
Methanol [FLAMMABLE]

Equipment
Round-bottomed flask (250 mL)
Reflux condenser and Dean–Stark trap
Heating/stirring mantle and stirrer bar
Heat gun
Rotary evaporator
Glassware for work-up

Note: Work wherever possible in a fume hood.

Place crude 2,6-bis(hydroxymethyl)-4-*t*-butylphenol, **27**, (20 g, 95 mmol) and *o*-xylene (150 mL) in a 250 mL round-bottomed flask fitted with a Dean–Stark trap and heat to reflux under nitrogen. It is worthwhile filling up the Dean–Stark trap with *o*-xylene prior to reflux as this will keep the volume of solvent in the reaction vessel constant during the experiment. Water generated condenses into the Dean–Stark trap during the course of the reaction. After 4 h, allow the mixture to cool and remove the solvent under reduced pressure to give an oily residue. Dissolve the residue in a minimum of warm dichloromethane (using a heat gun to effect dissolution) and add methanol until the solution starts to become cloudy. At this point a small volume of dichloromethane should be added to regenerate the clear solution. The product forms as a fine precipitate over 24 h and is isolated by filtration. 4-*t*-Butyloxacalix[3]arene (**28**) is obtained as a white powder.

Yield: 2.8 g (15%, subsequent crops contain increasing levels of linear polymers); m.p.: 240–242 °C; IR (v, cm^{-1}): 3370, 2960, 2870, 1485, 1210, 1080, 880; ^{1}H NMR (δ, ppm; CDCl$_3$): 8.6 (s, 3 H, O*H*), 6.9 (s, 6 H, Ar*H*), 4.7 (s, 6 H, OC*H$_2$*Ar), 1.3 (s, 27 H, C*H$_3$*).

2,6-Bis(hydroxymethyl)-4-phenylphenol (29)

Reagents
4-Phenylphenol
Sodium hydroxide [CORROSIVE]

Equipment
Round-bottomed flask (1 L)
Rotary evaporator

Aqueous formaldehyde (37%)
Tetrahydrofuran (THF) [FLAMMABLE]
Distilled water
Propan-2-ol [FLAMMABLE]
Acetone [FLAMMABLE]
Glacial acetic acid [CORROSIVE]
Toluene [FLAMMABLE]

Glassware for filtration and recrystallization

Note: Work wherever possible in a fume hood.

Dissolve 4-phenylphenol (42.6 g, 0.25 mol) in THF (50 mL) and place in a 1 L flask. Add a solution of sodium hydroxide (10.5 g in 50 mL of water) and, after cooling the mixture, follow this with 37 per cent formaldehyde (60 mL, 0.75 mol, 50 per cent excess). Leave the mixture for seven days at room temperature then remove the remaining THF by rotary evaporator to give a thick off-white paste. Add sufficient propan-2-ol (*ca.* 300 mL) to dissolve the paste and precipitate out the sodium 2,6-bis(hydroxymethyl)-4-phenylphenolate. Filter off the precipitate, weigh it (approximately 25 g should be formed) and suspended it in acetone (200 mL). Acidify with a stoichiometric amount of glacial acetic acid (0.25 mL per g of the sodium salt) in acetone (100 mL) to precipitate sodium acetate. Remove the latter by filtration then remove the remaining solvent under reduced pressure to give an off-white solid. Recrystallize if necessary from toluene to give acid-free 2,6-bis(hydroxymethyl)-4-phenylphenol (**29**) as a white crystalline solid.

Yield: *ca.* 25 g (40%); m.p.: 110–111 °C; IR (v, cm^{-1}): 3400, 1200, 1070, 1005; ^{1}H NMR (δ, ppm; CDCl$_3$): 8.2 (br s, 1 H, ArO*H*), 7.5 (d, 2 H, Ar*H*), 7.4 (t, 2 H, Ar*H*), 7.3 (s, 2 H, Ar*H*), 7.3–7.2 (m, 1 H, Ar*H*), 4.9 (s, 4 H, C*H*$_2$).

4-Phenyloxacalix[3]arene (30)

Reagents
2,6-Bis(hydroxymethyl)-4-phenylphenol (**29**)
o-Xylene [FLAMMABLE]
Diethyl ether [FLAMMABLE]
Dichloromethane [TOXIC]
Methanol [FLAMMABLE]

Equipment
Two-necked round-bottomed flask (250 mL)
Reflux condenser and Dean–Stark trap
Heating/stirring mantle and stirrer bar
Rotary evaporator
Glassware for precipitation and filtration

Note: Work wherever possible in a fume hood.

Place a mixture of crude 2,6-bis(hydroxymethyl)-4-phenylphenol, **29**, (15 g, 70 mmol) in *o*-xylene (125 mL) in a 250 mL round-bottomed flask fitted with a Dean–Stark trap and heat to reflux. It is worthwhile filling up the Dean–Stark trap with *o*-xylene prior to reflux as this will keep the volume of solvent in the reaction vessel constant during the experiment. Water generated is removed by the trap during the course of the reaction. A thick paste forms after *ca.* 2 h. After 4 h the mixture is allowed to cool and the solvent removed to give an oily residue. Add diethyl ether (20 mL) to the residue and remove under reduced pressure. Repeat this process twice more to azeotrope any remaining *o*-xylene. Add methanol (50 mL) and heat to reflux then filter to isolate a mixture of polymers and cyclic products. Add dichloromethane (50 mL) to the precipitate, heat to reflux and filter while hot. Wash the precipitated solids with dichloromethane (20 mL), combine the organic filtrates and remove the solvent under reduced pressure to leave a yellow solid (*ca.* 2 g). Dissolve in a minimum of dichloromethane then add methanol until the solution becomes cloudy. Add further dichloromethane drop-wise until a clear solution reforms and leave overnight at room temperature. Filter the resulting precipitate and wash with methanol (20 mL) to yield 4-phenylox-acalix[3]arene (**30**) as a pale yellow powder.

Yield: 0.53 g (*ca.* 4%); m.p.: 138–140 °C; IR (v, cm^{-1}): 3330, 3015, 2880, 1605, 1195, 1060, 885, 760, 700; ^{1}H NMR (δ, ppm; CDCl$_3$,): 10.5 (s, 3 H, O*H*), 7.6–7.3 (m, 21 H, Ar*H*), 3.5 (s, 12 H, OC*H*$_2$Ar).

3.6 Oxacalixarene Derivatives

As with the calix[n]arene family, the oxacalixarenes may also be derivatized to yield compounds with deeper cavities or greater functionality. Unfortunately, unlike the calix[n]arenes, the 4-*t*-butyl group cannot be removed using retro-Friedel–Crafts methods as the etheric links are not robust under the conditions necessary. The best way to introduce functionality to the upper rim is to build it in at the start. This approach may be problematic as the precursor 2,6-bis(hydroxy) phenol derivatives are often formed in low yield or are impossible to prepare in a single step. Despite these difficulties a diverse range of such compounds exist in addition to the simple 4-*t*-butyl and 4-methyl derivatives: bromo- and benzylox-acalix[3]arenes that bind fullerenes [1–3] have both been reported.

Lower rim derivatization is far easier to effect and examples include alkyl-, ester-, amide-[4], pyridyl- [5] and pyrene-appended [6] systems, not to mention the more exotic species that bind gallium [7] or give a colorimetric response when complexing chiral guests [8] as shown in Figure 3.19. Further examples include oxacalix[3]arenes that form capsules [9,10], palladium cross-linked dimers [11–13] and 2:1 inclusion complexes with C$_{60}$ [14,15]. Oxacalixarenes with lower rim substituents have even been incorporated into electrochemical sensors for

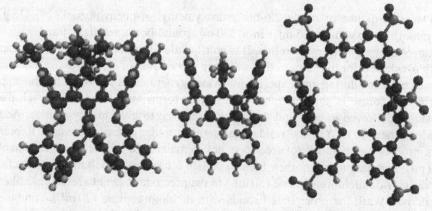

Figure 3.19 Upper and lower rim derivatives of oxacalix[3]arenes

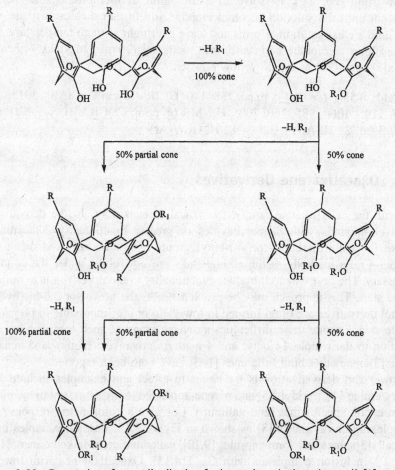

Figure 3.20 Expected conformer distribution for lower rim substituted oxacalix[3]arenes

dopamine [16]. In all these examples the configuration of the macrocycle is important. Oxacalix[3]arenes have a large enough cavity to allow 'through-the-annulus' rotation. The problem with this is that statistically only 25 per cent of the product should be in the *cone* conformer, as illustrated in Figure 3.20, which is very inefficient synthesis if a macrocyclic cavity is desired. Some applications of oxacalix[3]arenes actually require conformer switching, which is possible if one phenolic ring is unsubstituted or has only been methylated [8], but most need a stable cone to form. One route which increases the likelihood of *cone* formation is to prepare the *N,N*-diethylacetamide derivative [17] and subsequently cleave the amides to form the tricarboxylic acid [18].

For reasons that have yet to be fully determined, the cone form of this amide derivative can be isolated in greater than statistical amounts [19]. Indeed, one author [18] claims to prepare it in 90 per cent yield! The use of sodium hydride as the deprotonating agent may be the key factor. Sodium has been shown to bind between the phenolic oxygens and amide nitrogens [20]. It is likely that it acts as a template during lower rim substitution and is only removed when the final amide group is in place. One possible mechanism is shown in Figure 3.21. The same effect is not seen in the synthesis of the ester derivative [4].

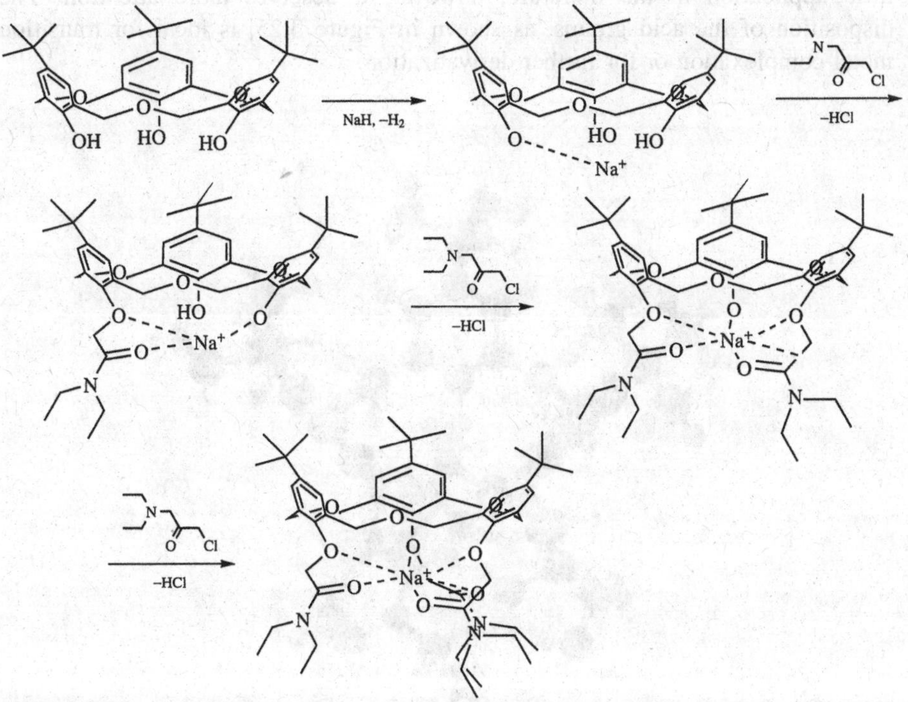

Figure 3.21 Possible template effect during the formation of 4-*t*-butyloxacalix[3]arene-tris(*N,N*-diethylacetamide)

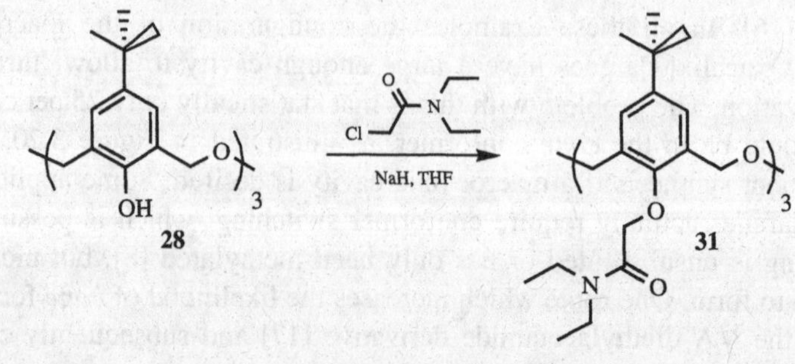

Figure 3.22 Synthesis of 4-*t*-butyloxacalix[3]arenetris(*N,N*-diethylacetamide) (**31**)

Here the synthesis of 4-*t*-butyloxacalix[3]arenetris(*N,N*-diethylacetamide) (**31**), based on the Shinkai method [17], is described and illustrated in Figures 3.22 and 3.23. The reaction between 4-*t*-butyloxacalix[3]arene and *N,N*-diethylchloroacetamide is, if anything, cleaner than that of the analogous 4-*t*-butylcalix[4]arene. The subsequent cleavage to form the tricarboxylic acid (**32**), as described by Yamato [18], is shown in Figure 3.24. The tricarboxylic acid derivative has found little application in the literature, however, it deserves more attention. The disposition of the acid groups, as shown in Figure 3.25, is ideal for transition metal complexation or for further derivatization.

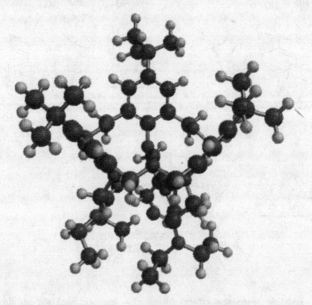

Figure 3.23 Structure of 4-*t*-butyloxacalix[3]arenetris(*N,N*-diethylacetamide)

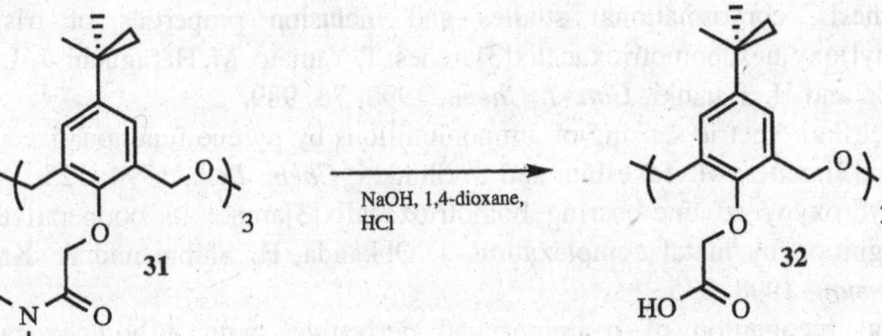

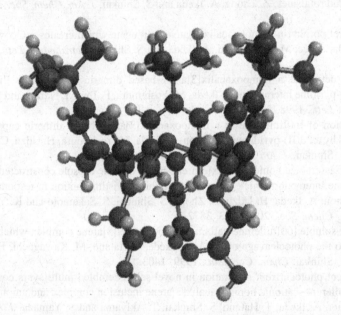

Figure 3.24 Synthesis of 4-*t*-butyloxacalix[3]arenetriacetic acid (**32**)

Figure 3.25 Simulated structure of 4-*t*-butyloxacalix[3]arenetriacetic acid

[1] Complexation of C_{60} with hexahomooxacalix[3]arenes and supramolecular structures of complexes in the solid state, K. Tsubaki, K. Tanaka, T. Kinoshita and K. Fuji, *Chem. Commun.*, 1998, 895.

[2] Preferential precipitation of C_{70} over C_{60} with *p*-halohomooxacalixarenes, N. Komatsu, *Org. Biomol. Chem.*, 2003, **1**, 204.

[3] A novel [60]fullerene–calixarene conjugate which facilitates self-inclusion of the [60]fullerene moiety into the homooxacalix[3]arene cavity, A. Ikeda, S. Nobukuni, H. Udzu, Z. L. Zhong and S. Shinkai, *Eur. J. Org. Chem.*, 2000, 3287.

[4] Conformational isomerism in and binding-properties to alkali-metals and an ammonium salt of *O*-alkylated homooxacalix[3]arenes, K. Araki, K. Inada, H. Otsuka and S. Shinkai, *Tetrahedron*, 1993, **49**, 9465.

[5] Synthesis, conformational studies and inclusion properties of tris[(2-pyridyl-methyl)oxy]hexahomotrioxacalix[3]arenes, T. Yamato, M. Haraguchi, J.-I. Nishikawa, S. Ide and H. Tsuzuki, *Can. J. Chem.*, 1998, **76**, 989.

[6] Novel fluorometric sensing of ammonium ions by pyrene functionalized homotrioxacalix[3]arenes, M. Takeshita and S. Shinkai, *Chem. Lett.*, 1994, 125.

[7] *N*-Hydroxypyrazinone-bearing homotrioxacalix[3]arene: its cooperative molecular recognition by metal complexation, J. Ohkanda, H. Shibui and A. Katoh, *Chem. Commun.*, 1998, 375.

[8] Chiral recognition of α-amino acid derivatives with a homooxacalix[3]arene: construction of a pseudo-C_2-symmetrical compound from a C_3-symmetrical macrocycle, K. Araki, K. Inada and S. Shinkai, *Angew. Chem. Int. Ed. Engl.*, 1996, **35**, 72.

[9] Triple linkage of two homooxacalix[3]arenes creates capsular molecules and self-threaded rotaxanes, Z. Zhong, A. Ikeda and S. Shinkai, *J. Am. Chem. Soc.*, 1999, **121**, 11906.

[10] A novel porphyrin-homooxacalix[3]arene conjugate which creates a C_3-symmetrical capsular space, M. Kawaguchi, A. Ikeda and S. Shinkai, *Tetrahedron Lett.*, 2001, **42**, 3725.

[11] Construction of a homooxacalix[3]arene-based dimeric capsule crosslinked by a Pd(II)-pyridine interaction, A. Ikeda, M. Yoshimura, F. Tani, Y. Naruta and S. Shinkai, *Chem. Lett.*, 1998, 587.

[12] Inclusion of [60]fullerene in a homooxacalix[3]arene-based dimeric capsule crosslinked by a Pd(II)-pyridine interaction, A. Ikeda, M. Yoshimura, H. Udzu, C. Fukuhara and S. Shinkai, *J. Am. Chem. Soc.*, 1999, **121**, 4296.

[13] A self-assembled homooxacalix[3]arene-based dimeric capsule constructed by a Pd^{II}-pyridine interaction which shows a novel chiral twisting motion in response to guest inclusion, A. Ikeda, H. Udzu, Z. Zhong, S. Shinkai, S. Sakamoto and K. Yamaguchi, *J. Am. Chem. Soc.*, 2001, **123**, 3872.

[14] Water-soluble [60]fullerene-cationic homooxacalix[3]arene complex which is applicable to the photocleavage of DNA, A. Ikeda, T. Hatano, M. Kawaguchi, H. Suenaga and S. Shinkai, *Chem. Commun.*, 1999, 1403.

[15] Efficient photocurrrent generation in novel self-assembled multilayers comprised of [60]fullerene–cationic homooxacalix[3]arene inclusion complex and anion porphyrin polymer, A. Ikeda, T. Hatano, S. Shinkai, T. Akiyama and S. Yamada, *J. Am. Chem. Soc.*, 2001, **123**, 4855.

[16] Dopamine-selective response in membrane potential by homooxacalix[3]arene triether host incorporated in PVC liquid membrane, K. Odashima, K. Yagi, K. Toda and Y. Umezawa, *Bioorg. Med. Chem. Lett.*, 1999, **9**, 2375.

[17] Syntheses and ion selectivities of the tri-amide derivatives of hexahomotrioxacalix[3]arene. Remarkably large metal template effect on the ratio of cone vs. partial-cone conformers. H. Matsumoto, S. Nishio, M. Takeshita and S. Shinkai, *Tetrahedron*, 1995, **51**, 4647.

[18] Synthesis and inclusion properties of C_3-symmetrically capped hexahomotrioxacalix[3]arenes with ester groups on the lower rim, T. Yamato, F. Zhang, H. Tsuzuki and Y. Miura, *Eur. J. Org. Chem.*, 2001, 1069.

[19] Conformational preferences of *O*-substituted oxacalix[3]arenes, P. J. Cragg, M. G. B. Drew and J. W. Steed, *Supramol. Chem.*, 1999, **11**, 5.

[20] A 'toothpaste tube' model for ion transport through trans-membrane channels, P. J. Cragg, M. C. Allen and J. W. Steed, *Chem. Commun.*, 1999, 553.

Derivatization of oxacalix[3]arenes

4-t-Butyloxacalix[3]arenetris(N,N-*diethylacetamide*) *(31)*

Reagents
4-*t*-Butyloxacalix[3]arene (**29**)
N,N-Diethylchloroacetamide
  [TOXIC; LACHRYMATOR]
Tetrahydrofuran (THF) [FLAMMABLE]
Dimethylformamide (DMF)
Sodium hydride (60% dispersion
  in mineral oil) [CORROSIVE; REACTS
  VIOLENTLY WITH WATER]
Distilled water
Hydrochloric acid (1 M) [CORROSIVE]
Diethyl ether [FLAMMABLE]

Equipment
Two-necked round-bottomed flask
  (250 mL)
Heater/stirrer and stirrer bar
Reflux condenser
Glass syringe and septum
Rotary evaporator
Glassware for work-up

Note: Work wherever possible in a fume hood.

Add 4-*t*-butyloxacalix[3]arene, **29**, (1 g, 1.55 mmol) and dry THF* (25 mL) to a two-necked round-bottomed flask charged with a magnetic stirrer bar and fitted with a reflux condenser. Stir at room temperature until the solids dissolve. Carefully add sodium hydride (0.40 g, 10.0 mmol [60 per cent dispersion in mineral oil]) to the stirred solution a little at a time through a glass funnel placed in the flask's second neck. Once all the solid has been added and the solution has stopped effervescing, rinse any remaining solids into the flask by pouring additional dry THF (25 mL) through the funnel. Remove the funnel and stopper the neck of the flask with a rubber septum. Carefully add 2-chloro-*N,N*-diethyl-acetamide (1.41 g, 1.29 mL, 9.38 mmol) by syringe through the septum to the stirred solution. When the addition is complete remove the septum and replace with a secure glass stopper. Reflux for 4 h then cool to room temperature and add water (30 mL). Acidify the sticky residue to pH 1 with hydrochloric acid (4 M solution) and extract with CH_2Cl_2 (3 × 25 mL). Wash the organic phase with water (30 mL) before removing solvent under vacuum. Add diethyl ether (50 mL) to the resulting white solid, filter and wash with more diethyl ether. Air dry to obtain 4-*t*-butyloxacalix[3]arenetris(*N,N*-diethylacetamide) (**31**) as a white powder.

Yield: 0.64 g (45%); m.p.: 215–217 °C; IR (v, cm^{-1}): 3400, 2965, 1655, 1485, 1460, 1200, 1095, 1075, 1060; ^{1}H NMR (δ, ppm; CDCl$_3$): 7.0 (s, 6 H, Ar*H*), 4.9

(d, 6 H, OCH_2Ar), 4.65 (d, 6 H, OCH_2Ar), 4.5 (s, 6 H, OCH_2CO), 3.4–3.3 (m, 12 H, NCH_2), 1.2–1.15 (m, 9 H, NCH$_2$CH_3), 1.1–1.05 (s, 27 H, CH_3).

*See compound 1

4-t-Butyloxacalix[3]arenetriacetic acid (32)

Reagents
4-t-Butyloxacalix[3]arene-
  (N,N-diethylacetamide) (31)
1,4-Dioxane [FLAMMABLE]
Sodium hydroxide (1 M solution)
  [CORROSIVE]
Hydrochloric acid (2 M solution)
  [CORROSIVE]
Ethyl acetate [FLAMMABLE]
Distilled water
Aqueous sodium chloride (saturated)
Sodium sulphate
Diethyl ether [FLAMMABLE]
Methanol [FLAMMABLE]

Equipment
Round-bottomed flask (100 mL)
Magnetic stirrer/heater
  and stirrer bar
Reflux condenser
Glassware for work-up

Note: Work wherever possible in a fume hood.

Add 4-butyloxacalix[3]arenetris(N,N-diethylacetamide), 31, (1 g, 1.1 mmol) and 1,4-dioxane (30 mL) to a single-necked 100 mL round-bottomed flask containing a magnetic stirrer bar. Stir the mixture and add aqueous sodium hydroxide (30 mL, 1 M solution). Fit the flask with a condenser and reflux for 72 h ensuring that the reaction never runs dry through the addition of extra 1,4-dioxane, if necessary. Concentrate the cooled solution by removing the 1,4-dioxane under reduced pressure then acidify to pH 1 with hydrochloric acid (2 M solution). Extract the resulting dispersion with ethyl acetate (2 × 20 mL) then wash the combined extracts with water (2 × 20 mL) and saturated aqueous sodium chloride (20 mL). Dry the ethyl acetate phase over sodium sulphate, filter and remove the solvent under reduced pressure. Wash the residue with diethyl ether and recrystallize from methanol (*ca.* 10 mL). 4-Butyloxacalix[3]arenetriacetic acid (32) is isolated as a white powder.

Yield: 0.4 g (40%); m.p.: 228 °C; IR (v, cm^{-1}): 3400, 2975, 2870, 1760, 1455, 1360, 1200, 1095; ^{1}H NMR (δ, ppm; CDCl$_3$): 6.95 (s, 6 H, ArH), 4.9 (d, 6 H, ArCH_2O), 4.65 (s, 6 H, ArOCH_2), 4.4 (d, 6 H, ArCH_2O), 1.1 (s, 27 H, CCH_3).

3.7 Azacalix[3]arenes

Analogous to the homooxacalixarenes described in the previous sections, the homoazacalix[3]arenes contain nitrogen bridgeheads that may incorporate an array of substituents. Azacalix[3]arenes are synthesised either through acid-catalysed cyclization of 2,6-bis(hydroxymethyl)phenols with amines, followed by thermal dehydration, or base-catalysed cyclization of 2,6-di(halomethyl)phenols with amines.

The first reported azacalixarene synthesis was by Hultsch [1] who prepared a direct analogue of oxacalixarene, hexahomotriazacalix[3]arene, through reaction of 4-t-butylphenol with hexamethylenetetramine, however, his method has yet to be verified. The first practical synthesis of an azacalix[3]arene was that of 4-methyl-(N-benzyl)triazacalix[3]arene and its higher homologues by Takemura [2] and a variation of this is described below. A convergent synthesis has since been developed by Hampton to give the compound in 95 per cent yield [3].

An alternative to this method, and one by which a variety of N-substituents may be introduced, is through the base catalysed reaction of a suitable amine such as an amino acid ester with 2,6-bis(chloromethyl)phenols [4]. The advantage of this method is that the N-substituents may be derived from volatile amines which would not be stable in refluxing toluene.

If aza- and oxacalix[3]arenes are prepared in the presence of an organic acid with complementary symmetry to the calixarene to be formed there appears to be no evidence for the formation of tetramers or higher homologues. The synthesis given here from the appropriate bis(hydroxymethyl)phenol precursor (33), shown in Figure 3.26, is of 4-methyl-(N-benzyl)triazacalix[3]arene (34) using Takemura's method but including an organic acid, p-toluenesulphonic acid, to template the formation of the trimeric species. A simulation of the product is shown in Figure 3.27.

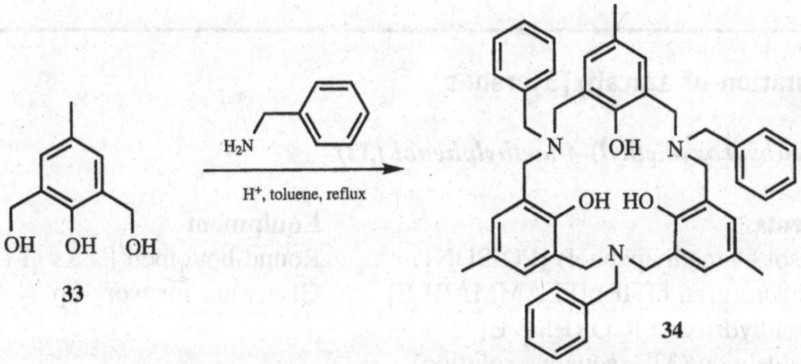

Figure 3.26 Synthesis of 4-methyl-(N-benzyl)azacalix[3]arene (34)

Figure 3.27   Simulated structure of 4-methyl-(*N*-benzyl)azacalix[3]arene

[1] Ring-Kondensate in Alkylphenolharzen, K. Hultzsch, *Kunststoffe*, 1962, **52**, 19.
[2] The first synthesis and properties of hexahomotriazacalix[3]arene, H. Takemura, K. Yoshimura, I. U. Khan, T. Shinmyozu and T. Inazu, *Tetrahedron Lett.*, 1992, **33**, 5775.
[3] A convergent synthesis of hexahomotriazacalix[3]arene macrocycles, P. Chirakul, P. D. Hampton and Z. Bencze, *J. Org. Chem.*, 2000, **65**, 8297.
[4] Synthesis, X-ray structure and alkali-metal binding properties of a new hexahomtriazacalix[3]arene, P. D. Hampton, W. Tong, S. Wu and E. N. Duesler, *J. Chem. Soc., Perkin Trans. 2*, 1996, 1127.

Preparation of azacalix[3]arenes

2,6-Bis(hydroxymethyl)-4-methylphenol (33)

Reagents
p-Cresol (4-methylphenol) [POISON]
Tetrahydrofuran (THF) [FLAMMABLE]
Sodium hydroxide [CORRSIVE]
Formaldehyde (37% aqueous solution)
  [TOXIC; CARCINOGEN]

Equipment
Round-bottomed flasks (1 L)
Glassware for work-up

Propan-2-ol [FLAMMABLE]
Glacial acetic acid [CORROSIVE]
Acetone [FLAMMABLE]
Toluene [FLAMMABLE]

Note: Work wherever possible in a fume hood.

Dissolve *p*-cresol (27 g, 25 mmol) in THF (55 mL) in a 1 L flask. Add a solution of sodium hydroxide (10 g in 30 mL of water) and, after cooling the mixture, follow this with 37 per cent formaldehyde (45 mL, 55 mmol, 10 per cent excess). Leave the mixture for 7 days at room temperature, by which time it will have solidified. If it has not, remove the remaining THF by rotary evaporator to give a thick off-white paste. Add sufficient propan-2-ol (*ca.* 300 mL) to free up the precipitate of sodium 2,6-bis(hydroxymethyl)-4-methylphenolate. Filter the precipitate, weigh it and suspend it in acetone (200 mL). Acidify with a stoichiometric amount of glacial acetic acid (0.25 mL per g of sodium salt) in acetone (100 mL) to precipitate sodium acetate, which can be removed by filtration. Remove the remaining solvent under reduced pressure to give an off-white solid. Recrystallization of the solid from toluene (100 mL) gives the acid-free 2,6-bis(hydroxymethyl)-4-methylphenol (**33**) as a fine powder.

Yield: 9.4 g (22%); m.p.: 130–131 °C; IR (v, cm^{-1}): 3388, 3304, 2950, 2911, 2876, 1483, 1464, 1246, 1060, 1004; ^{1}H NMR (δ, ppm; CDCl$_3$/DMSO): 6.8 (s, 2 H, Ar*H*), 4.5 (s, 4H, C*H*$_2$), 2.1 (s, 3H, C*H*$_3$).

4-Methyl(N-benzyl)azacalix[3]arene (34)

Reagents	**Equipment**
2,6-Bis(hydroxymethyl)-4-methylphenol (**33**)	Round-bottomed flask (250 mL)
Benzylamine [CORROSIVE]	Heater/stirrer and stirrer bar
p-Toluenesulphonic acid [CORROSIVE]	Dean–Stark trap
Toluene [FLAMMABLE]	Condenser
Methanol [FLAMMABLE]	Glassware for filtration
Acetone [FLAMMABLE]	and chromatography
Silica	
Dichloromethane [TOXIC]	

Note: Work wherever possible in a fume hood.

Place a mixture of 2,6-bis(hydroxymethyl)-4-methylphenol, **33**, (4.0 g, 24 mmol), benzylamine (2.57 mL, 24 mmol), and *p*-toluenesulphonic acid (1.52 g, 8 mmol) in toluene (150 mL) in a 250 mL round-bottomed flask fitted with a Dean–Stark trap, filled with toluene, and reflux for 24 h. Add more toluene to the reaction flask as necessary if the solvent appears to be evaporating. This occasionally occurs when the reaction is set up in a powerful fume hood. Water generated is removed by the Dean–Stark trap during the course of the reaction. After 24 h, cool the mixture and filter off any solids, washing with dichloromethane (25 mL), then remove the solvent under reduced pressure to give lime green glass. Dissolve in a minimum of dichloromethane (ca. 20 mL). Isolate the product following column chromatography (silica, dichloromethane:methanol 9:1). After checking purity by thin layer chromatography (aluminium-backed silica, dichloromethane:methanol 9:1) combine the bright yellow fractions and remove the solvents under reduced pressure to leave 4-methyl(N-benzyl)azacalix[3]arene (**34**) as a yellow solid.

Yield: 2.6 g (45%); m.p.: 119–121 °C; IR (v, cm^{-1}): 3026, 2913, 2799, 1603, 1480, 1310; ^{1}H NMR (δ, ppm; CDCl$_3$): 10.7 (br s, 3 H, OH), 7.4–7.1 (m, 15 H, PhH), 6.7 (s, 6 H, ArH), 3.7 (s, 12 H, ArCH_2), 3.6 (s, 6 H, CH_2Ph), 2.15 (s, 9 H, CH_3).

3.8 Calixarene Variations

The preceding 'molecular baskets' belong broadly to two classic classes of macrocycles namely the calixarenes and cycloveratrylenes. There are several other macrocycles with the potential to bind guests in an aromatic-rich cavity that are worthy of discussion. Of the vast amount of work that emanated from the Cram group, the rigid spherands and carcerands stand out. In addition there are the cyclophanes and Baeyer's early contribution to macrocyclic chemistry, resorcinarenes and calixpyrroles.

Cyclophanes encompass the entire range of macrocycles comprised of linked aromatic groups; the definition therefore also includes the calixarenes, cyclotriveratrylenes and many other related systems. Despite this, most supramolecular chemists consider cyclophanes to be those macrocycles in which aromatic rings are joined in the 1,4-positions by short alkyl bridges. The original cyclophane, reported in 1899, consists of two benzene rings linked through the 1,3-positions by ethyl bridges and is named, according to cyclophane terminology, [2.2]metacyclophane [1]. The number of atoms in each bridge is given in square brackets and is followed by the position of aromatic substitution (ortho-, meta- or para-) and the non-systematic name 'cyclophane'. It took 50 years for rational synthesis to appear with Brown and Farthing isolating small amounts of [2,2]paracyclophane (where the rings are linked in the 1,4-positions) in 1949 [2]. This was followed in 1951 by Cram and Steinberg's [2,2]paracyclophane synthesis which ushered in a new era for this class of compounds [3]. Several methods can now be used to prepare cyclophanes including sulphur extrusion (aromatic groups are linked through thioether bridges that are later pyrolysed [4]) and Wurtz coupling [5].

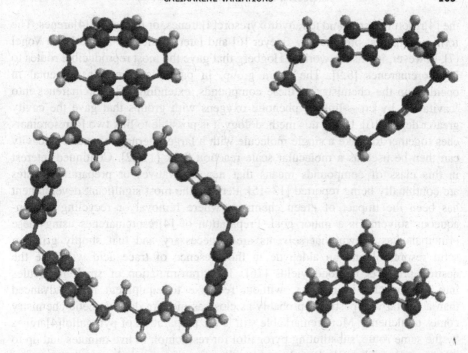

Figure 3.28 Some representative cyclophanes: [2.2]paracyclophane (top left), [3.3.1] para-cyclophane (top right), an azacyclophane (bottom left), [3.3]metacyclophane (bottom right)

Examples of cyclophanes that fit a more traditional description are illustrated in Figure 3.28.

Some of the most remarkable advances in recent years have been the improved synthetic routes to the related cyclophane-like compounds shown in Figure 3.29,

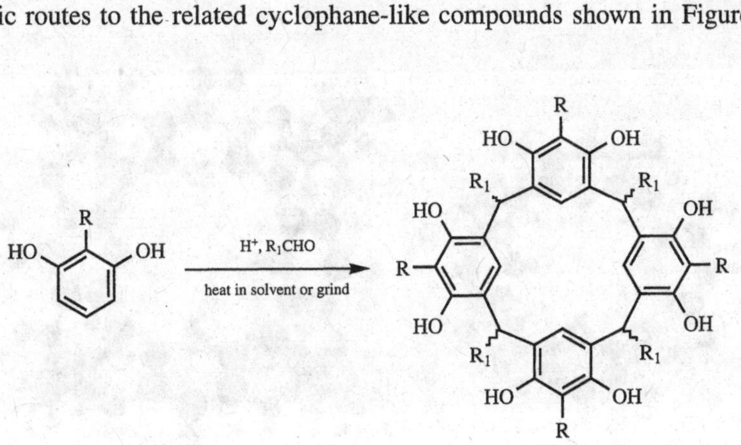

Figure 3.29 Synthesis of [4]resorcinarenes (**35**: R = H, R_1 = CH$_3$) and pyrogallol[4]arenes (**36**: R = OH, R_1 = (CH$_2$)$_6$CH$_3$)

the [4]resorcinarenes and tetrahydroxyresorc[4]arenes, or pyrogallol[4]arenes. The former were first prepared by Baeyer [6] and later studied by Niederl and Vogel [7], however, it was the work of Högberg that gave the most reproducible routes to [4]resorcinarenes [8,9]. The Cram group, in particular, were instrumental in opening up the chemistry of these compounds, extending [4]resorcinarenes into 'cavitands' by cross-linking phenolic oxygens with groups that gave the cavity greater depth [10]. Using this methodology it is possible to link two [4]resorcinarenes together to make a single molecule with a large internal volume. The cavity can then be used as a molecular scale reaction flask [11,12]. Continued interest in this class of compounds means that new derivatives or preparative routes are continually being reported [13–15]. Perhaps the most significant development has been the impact of green chemistry where removal or recycling of non-aqueous solvents is a major goal. Preparation of [4]resorcinarenes using these principals has shown that solvents are unnecessary and that simply grinding solid resorcinol and an aldehyde in the presence of trace acid will give the desired products in good yields [16]. This transmutation of small molecules into macrocycles in minutes, without recourse to equipment more advanced than a mortar and pestle, is probably as close as modern day synthetic chemistry comes to alchemy. More remarkable still is the preparation of pyrogallol[4]arenes by the same route, substituting pyrogallol for resorcinol, in five minutes and up to 90 per cent yield [17]. Simple recrystallization from ethyl acetate generates stable nanocapsules comprised of six macrocycles that contain seven solvent molecules.

The examples given here employ two routes: Högberg's original method and Mattay's [18] extended reflux method. The products are simple C-methyl[4]resorcinarene derivatives and a heptyl pyrogallol[4]arene as shown in Figure 3.30.

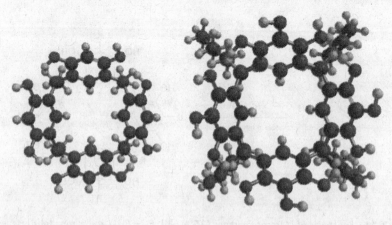

Figure 3.30   C-Methyl[4]resorcinarene (left) and C-heptylpyrogallol[4]arene (right)

Other linear aldehydes may be used to prepare pyrogallol[4]arenes though some, such as pentanal, have unpleasant odours.

[1] Contribution à l'étude de la réaction de Fittig, M. M. Pellegrin, *Recl. Trav. Chim. Pays-Bas*, 1899, **18**, 457.

[2] Preparation and structure of di-*p*-xylene, C. J. Brown and A. C. Farthing, *Nature*, 1949, **164**, 915.

[3] Macro rings I. Preparation and spectra of the paracyclophanes, D. J. Cram and H. Steinberg, *J. Am. Chem. Soc.*, 1951, **73**, 5691.

[4] Metacyclophanes and related compounds 1. Preparation and nuclear magnetic resonance spectra of 8,16-disubstituted [2.2]metacyclophanes, M. Tashiro and T. Yamato, *J. Org. Chem.*, 1981, **46**, 1541.

[5] The chemistry of xylylenes VI. The polymerization of *p*-xylylene, L. A. Errede, R. S. Gregorian and J. M. Hoyt, *J. Am. Chem. Soc.*, 1960, **82**, 5218.

[6] Uber die Verbindung dar Aldehyde mit den Phenolen, A. Baeyer, *Ber. Dtsch. Chem. Ges.*, 1872, **5**, 25.

[7] Aldehyde–resorcinol condensations, J. B. Niederl and H. Vogel, *J. Am. Chem. Soc.*, 1940, **62**, 2512.

[8] Stereoselective synthesis and DNMR study of two 1,8,15,22-tetraphenyl[1_4]metacyclophan-3,5,10,12,17,19,24,26-octols, A. G. S. Högberg, *J. Am. Chem. Soc.*, 1980, **102**, 6046.

[9] Two stereoisomeric macrocyclic resorcinol-acetaldehyde condensation products, A. G. S. Högberg, *J. Org. Chem.*, 1980, **45**, 4498.

[10] Host–guest complexation 48. Octol building blocks for cavitands and carcerands, L. M. Tunstad, J. A. Tucker, E. Dalcanale, J. Weiser, J. A. Bryant, J. C. Sherman, R. C. Helgeson, C. B. Knobler and D. J. Cram, *J. Org. Chem.*, 1989, **54**, 1306.

[11] The taming of cyclobutadiene, D. J. Cram, M. E. Tanner and R. Thomas, *Angew. Chem. Int. Ed. Engl.*, 1991, **30**, 1024.

[12] The phenylnitrene rearragement in the inner phase of a hemicarcerand, R. Warmuth and S. Makowiec, *J. Am. Chem. Soc.*, 2005, **127**, 1084.

[13] Bismuth compounds in organic synthesis. Synthesis of resorcinarenes using bismuth triflate, K. E. Peterson, R. C. Smith and R. S. Mohan, *Tetrahedron Lett.*, 2003, **44**, 7723.

[14] A novel synthesis of parent resorc[4]arene and its partial alkyl ethers, J. Stursa, H. Dvorakova, J. Smidrkal, H. Petrickova and J. Moravcova, *Tetrahedron Lett.*, 2004, **45**, 2043.

[15] A versatile six-component molecular capsule based on benign synthons – selective confinement of a heterogeneous molecular aggregate, G. W. V. Cave, M. J. Hardie, B. A. Roberts and C. L. Raston, *Eur. J. Org. Chem.*, 2001, 3227.

[16] Solvent-free synthesis of calix[4]resorcinarenes, B. A. Roberts, G. W. V. Cave, C. L. Raston and J. L. Scott, *Green Chem.*, 2001, **3**, 280.

[17] Inner core structure responds to communication between nano-capsule walls, G. W. V. Cave, J. Antesberger, L. J. Barbour, R. M. McKinlay and J. L. Atwood, *Angew. Chem. Int. Ed. Engl.*, 2004, **43**, 5263.

[18] Self-assembly of a 2,8,14,20-tetraisobutyl-5,11,17,23-tetrahydroresorc[4]arene, T. Gerkensmeier, W. Iwanek, C. Agena, R. Fröhlich, S. Kotila, C. Näther and J. Mattay, *Eur. J. Org. Chem.*, 1999, 2257.

Preparation of [4]resorcinarenes and pyrogallol[4]arenes

C-Methyl[4]resorcinarene (35)

Reagents

Resorcinol [HARMFUL]
Acetaldehyde [FLAMMABLE]
Distilled water
Hydrochloric acid (37%) [CORROSIVE]
Ethanol [FLAMMABLE]

Equipment

100 mL round-bottomed flask
Heater/stirrer and stirrer bar
Glassware for filtration
  and recrystallization

Note: Work wherever possible in a fume hood.

Dissolve resorcinol (5.5 g, 50 mmol) in water (20 mL) in a 100 mL round-bottomed flask and add hydrochloric acid (5 mL, 37 per cent aqueous solution) dropwise with vigorous stirring. To the stirred solution, add redistilled acetaldehyde (2.6 mL, 2.2 g, 50 mmol) dropwise. Heat to 75 °C for 1 h then cool in ice and filter of the beige precipitate. Wash with cold water (*ca.* 5 mL) and recrystallize from ethanol (*ca.* 35 mL) to give C-methyl[4]resorcinarene (35) as a white powder.

Yield: 3.9 g (50%); m.p.: >250 °C; IR (v, cm^{-1}): 3510, 3170, 2265, 1620, 1340, 975, 725; ^{1}H NMR (δ, ppm; CD$_3$CN): 7.75 (br s, 8 H, ArO*H*), 7.35 (s, 4 H, Ar*H*), 6.2 (s, 4 H, Ar*H*), 4.4 (q, 8 H, C*H*), 1.7 (m, 12 H, C*H*$_3$).

C-Heptylpyrogallol[4]arene (36)

Reagents

Pyrogallol
Octanal
Ethanol (96%) [FLAMMABLE]
Hydrochloric acid (37%) [CORROSIVE]
Acetonitrile [FLAMMABLE]

Equipment

100 mL round-bottomed flask
Ice bath
Heater/stirrer and stirrer bar
Inert atmosphere line
Glassware for filtration
  and recrystallization

Note: Work wherever possible in a fume hood.

Dissolve pyrogallol (5.0 g, 40 mmol) in ethanol (30 mL) in a 100 mL round-bottomed flask and cool in ice. Add hydrochloric acid (6 mL, 37 per cent aqueous solution) dropwise with vigorous stirring, keeping the temperature below 5 °C. To the stirred solution, add octanal (6.25 mL, 5.1 g, 40 mmol) dropwise, again keeping the temperature below 5 °C. Following the addition, reflux the colourless solution under an inert atmosphere for 18 h. As the reaction mixture warms it turns progressively deeper red. Once the reflux has completed and the purple mixture cooled to room temperature, filter the precipitate and wash with distilled water until the washings are colourless. Recrystallize the precipitate from acetonitrile (*ca.* 20 mL per g crude product) to give *C*-heptylpyrogallol[4]arene (**36**) as a microcrystalline black solid.

Yield: 3.3 g (85%); m.p.: >250 °C; IR (v, cm^{-1}): 3585, 3380, 1620, 1520, 1426, 1235, 1200, 1120, 1100, 1020, 995, 835; ^{1}H NMR (δ, ppm; CD$_3$CN): 7.4 (s, 8 H, ArO*H*), 6.85 (s, 4 H, ArO*H*), 6.3 (s, 4 H, Ar*H*), 4.2 (t, 8 H, C*H*), 1.5–1.1 (m, 42 H, C*H*$_2$), 0.9 (t, 12 H, C*H*$_3$).

3.9 Molecular Cages for Cations and Anions

The complexation of cations has been achieved by inorganic chemists for decades as shown by the copper phthalocyine complex in Chapter 2. Early success occurred when it was found that transition metals could template Schiff base condensations between primary amines and aldehydes. Metals preferring square planar geometries were particular good candidates: the nitrogen lone pairs present in diamines such as ethylenediamine, propylenediamine and *o*-phenylenediamine coordinate to metals and are then predisposed to react with dialdehydes able to bridge between two amine termini. Polyamines such as triethylenetetramine or *N*,*N'*-bis(2-amino-ethyl)-1,3-propanediamine are able to bind to all four coordination sites available to square planar transition metals. Addition of a dialdehyde completes the macrocycle's formation. Indeed, this is the method by which many tetraazamacro-cycles are prepared. Interest in these compounds has seen a recent resurgence with their potential medical uses, not least as anti-human immunodeficiency virus agents [1]. Nevertheless, all of these examples are essentially planar complexes.

Transition metals are bound by an array of functional organic groups in both natural and unnatural contexts. As targets for 'three-dimensional' complexation they would seem ideal: discrimination based on size, charge density, polarizability and preferred orbital orientation should allow the supramolecular chemist plenty of scope to design metal-specific ligands. In practice most transition metal complexants are nitrogen- and sulphur-containing ligands derived from cyclic and acyclic Schiff bases, as shown in Figure 3.31. A glance back at the molecules described in the previous chapters will show that a majority of the compounds described are derived from ethers or phenols. As such they are ill-disposed to bind transition metals and will more often be found complexing alkali metals or small neutral molecules. There are of course many ligands that can be prepared which contain the design elements necessary to entrap particular transition metals as evidenced by the number of books and journals dedicated to transition metal coordination chemistry. In order to encapsulate transition metals it is necessary to have ligands with convergent binding sites that are either templated around the metal or are flexible enough to allow a metal to enter the central void once the ligand has formed. The former approach has been used by Sargeson [2] who encapsulated

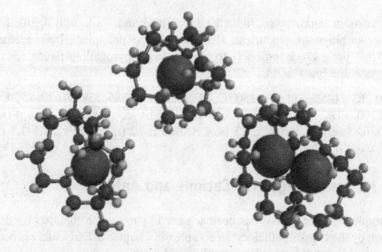

Figure 3.31 Encapsulating ligands for transition metals: Sargeson's sepulchrates (bottom left and top centre), Nelson's *tren*-cryptate (bottom right)

metals in sepulchrates, the latter was taken by Nelson to synthesize *tren*-derived cryptands to bind transition metals [3].

Perhaps the first examples of inclusion phenomena that do not involve traditional transition metal coordination compounds have been the crown ether complexes of alkali and alkaline earth cations. While these complexes were prepared by Pedersen in the late 1960s it was only through the application of single crystal X-ray techniques that the true nature of the inclusion phenomena was revealed. Early crystal structures of dibenzo[18]crown-6 with rubidium [4] showed the size–fit relationship between hosts and guests. Complementarity was related to the van der Waals radii of the cations and apparent cavity size by Christensen in 1971 [5]: a concept that has been central to supramolecular chemistry ever since. Metal binding by all-oxygen containing crown ethers and polyethers is limited to alkali and alkaline earths, transition metals with full d-shells (e.g. Ag^+ [6]), lanthanides and actinides. With the exception of examples where the uranyl ion, UO_2^+, threads through the centre of the crown, all involve metals in which there is no particular preferential angular dependence for orbital overlap with the donor atoms of the crown. As yet there are no examples of true crown ether inclusion complexes with transition metals in tetrahedral, square planar or octahedral geometries. Substitution of one or more oxygen atoms with sulphur, nitrogen or phosphorous creates an environment that is far more inviting for transition metals. This is particularly true of the [3]crown-9 analogues, [9]ane-S_3, [9]ane-N_3 and their derivatives [7,8], though the coordination complexes of the phosphorus analogue, [9]ane-P_3, are rather scarce [9].

The relationship between cavity size and van der Waals radii is reinforced when the complex is stabilized by additional factors, such as symmetry, that impose an alternating up–down pattern of donor atoms around the metal guest. Some of the size–fit issues have been explored by Hancock [10,11]. Although polyethers and related podands are more flexible, and thus may be expected to form weaker complexes, they may also organize around the guest. An excellent example of this is Hosseini's hexaethylene glycol isonicotine derivative. The polyether moiety is able to encircle a silver cation whilst the nicotinyl termini occupy axial positions on adjacent metal ions. When donor atoms are constrained within rigid planar macrocycles that force them to adopt alternating positions they can create an environment for cations while leaving axial sites open to further coordinating ligands as is seen in the alkali metal picrate complexes of torands [12]

Anions are, by their very nature, harder species to complex than cations or neutral molecules. Even the 'simple' anions such as the halides have the associated problems that they are relatively large compared to other spherical species. This necessitates the design of large cyclic or encapsulating ligands in order to engender strong and specific binding, as illustrated in Figure 3.32. Unfortunately, large macrocycles have high conformational mobility which must be overcome prior to guest inclusion. There is also the possibility of competitive binding by other, smaller species. For example, it is likely that a ligand designed to bind chloride, and containing polyamine or quaternary ammonium groups for that purpose, may well take in fluoride as well. With the latter's higher charge density it

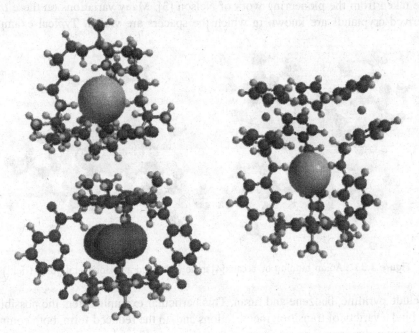

Figure 3.32 Encapsulating ligands for anions: Schmidtchen's chloride-binding ligand (top left), Anslyn's nitrate-binding cryptand (bottom left), Steed's chloride-detecting ferrocenyl podand (right)

is unlikely that it could be displaced by chloride. Of course not all anions are spherical. Design of ligands for carbonate and nitrate should be able to discriminate between the two based on charge and size, though not on geometry. A similar argument may be extended to the tetrahedral phosphate and sulphate anions [13]. Specific binding of more esoteric tetrahedral species such as pertechnetate, chromate, molybdate and permanganate is a greater challenge.

If protonated marocycles, or those carrying a permanent positive charge, are to be avoided other methods of inducing anions into the macrocyclic cavity must be found. One such approach to encapsulate tetrahedral anions, pioneered by Steed and Atwood, is to prepare organometallic derivatives of calixarenes and cyclotriveratrylenes. In these compounds the aromatic rings form part of an organometallic 'sandwich' compound where the metal and second aromatic group lie outside the macrocyclic cavity. When iron and ruthenium are used in this context they reduce the electron density of the macrocyclic aromatic ring making the cavity more inviting to electron rich species, particularly anions as shown in Figure 3.33.

Ligands based on tris(ethylamino)ethane, *tren*, have been discussed previously, as have azacryptands. A related class of ligands, *tren*-derived azacryptands, can be tuned to discriminate in favour of particular anions; for example, the distances between amine groups at either apex of an azacryptand can be varied to accommodate dicarboxylic acids of different lengths. The examples given here are taken from the pioneering work of Nelson [3]. Many variations on these *tren*-derived cryptands are known in which the spacers are varied. Typical examples

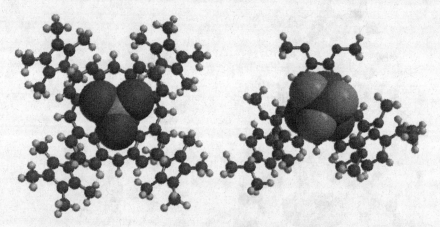

Figure 3.33 Anion binding by a calix[4]arene (left) and a cyclotriveratrylene (right)

include pyridine, benzene and furan. This particular example offers the possibility to bind a variety of transition metal cations and, in the reduced form, both common and unusual anions. The synthesis of the cryptand, **37**, and its reduction to form an anion-binding derivative, **38**, are shown in Figure 3.35. An example of the simulated structure of an anion complex is illustrated in Figure 3.36.

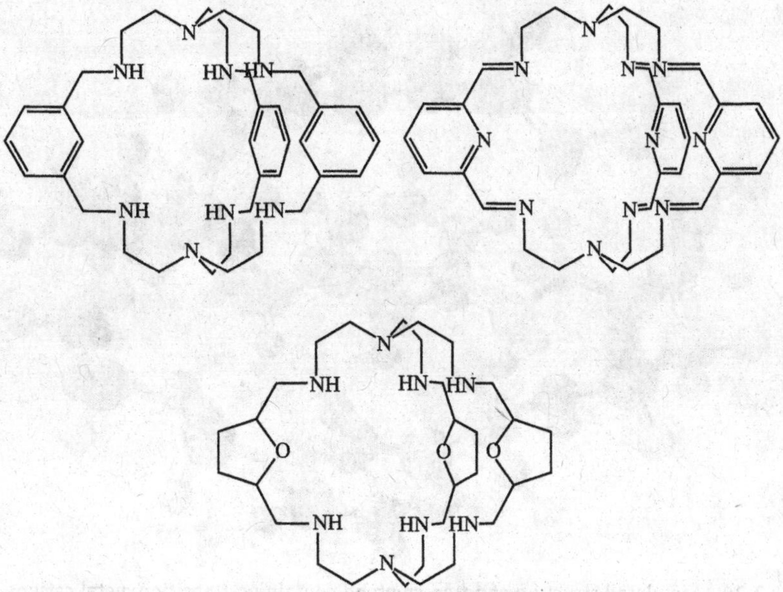

Figure 3.34 *Tren*-derived ligands

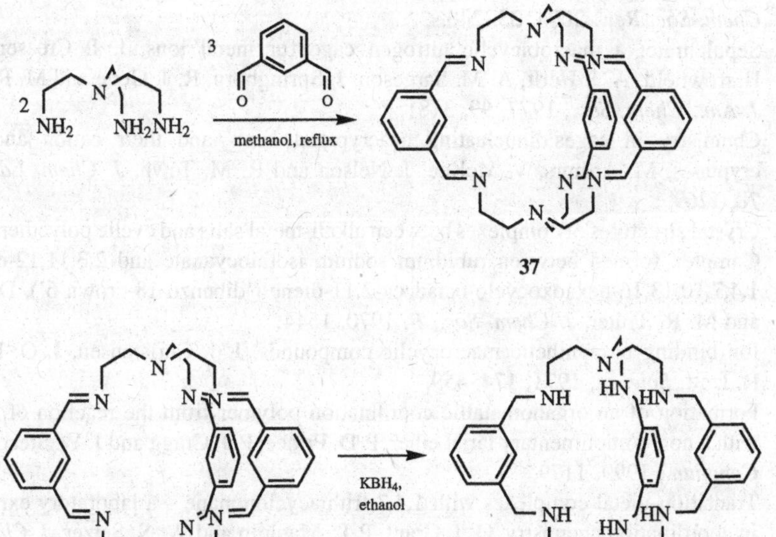

Figure 3.35 Synthesis of *tren*-cryptand (**37**) and its reduced form (**38**)

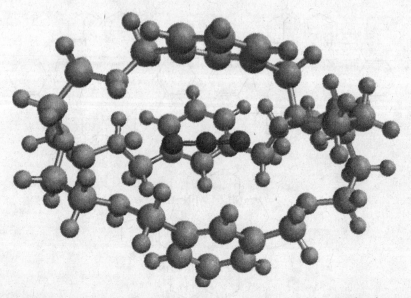

Figure 3.36 Simulated structure of a *tren*-cryptand containing transition metal cations and an azide anion

[1] Cyclam complexes and their applications in medicine, X. Y. Liang and P. J. Sadler, *Chem. Soc. Rev.*, 2004, **33**, 246.

[2] Sepulchrate: a macrobicyclic nitrogen cage for metal ions, I. I. Creaser, J. M. Harrowfield, A. J. Herlt, A. M. Sargeson, J. Springborg, R. J. Geue and M. R. Snow, *J. Am., Chem. Soc.*, 1977, **99**, 3181.

[3] Chemistry in cages:dinucleating azacryptand hosts and their cation and anion cryptates, M. Arthurs, V. McKee, J. Nelson and R. M. Town, *J. Chem. Ed.*, 2001, **78**, 1269.

[4] Crystal structures of complexes between alkali-metal salts and cyclic polyethers. Part I. Complex formed between rubidium sodium isothiocyanate and 2,3,11,12-dibenzo-1,4,7,10,13,16-hexaoxocyclo-octadeca-2,11-diene ('dibenzo-18-crown-6'), D. Bright and M. R. Truter, *J. Chem. Soc., B*, 1970, 1544.

[5] Ion binding by synthetic macrocyclic compounds, J. J. Christensen, J. O. Hill and R. Izatt, *Science*, 1971, **174**, 459.

[6] Formation of an organometallic coordination polymer from the reaction of silver(I) with a non-complimentary lariat ether, P. D. Prince, P. J. Cragg and J. W. Steed, *Chem. Commun.*, 1999, 1179.

[7] Transition-metal complexes with 1,4,7-trithiacyclononane – a laboratory experiment in coordination chemistry, G. J. Grant, P. L. Mauldin and W. N. Setzer, *J. Chem. Ed.*, 1991, **68**, 605.

[8] Synthetic and structural aspects of the chemistry of saturated polyaza macrocyclic ligands bearing pendant coordinating groups attached to nitrogen, K. P. Wainwright, *Coord. Chem. Rev.*, 1997, **166**, 35.

[9] Template synthesis of the first 1,4,7-triphosphacyclononane derivatives, P. G. Edwards, P. D. Newman and K. M. A. Malik, *Angew. Chem., Int. Ed.*, 2000, **39**, 2922.

[10] Macrocycles and their selectivity for metal-ions on the basis of size, R. D. Hancock, *Pure & Appl. Chem.*, 1986, **58**, 1445.

[11] Chelate ring size and metal-ion selection – the basis of selectivity for metal-ions in open-chain ligands and macrocycles, R. D. Hancock, *J. Chem. Ed.*, 1992, **69**, 615.

[12] Conformational preference of the torand ligand in its complexes with potassium and rubidium picrate, T. W. Bell, P. J. Cragg, M. G. B. Drew, A. Firestone and A. D.-I. Kwok, *Angew. Chem. Int. Ed. Engl.*, 1992, **31**, 345.

[13] Encapsulated sulfates: insight into binding propensities, S. O. Kang, M. A. Hossain, D. Powell and K. Bowman-James, *Chem. Commun.*, 2005, 328.

Preparation and reduction of *tren*-derived cryptands

Tren-*cryptand (37)*

Reagents
Tris(2-aminoethyl)amine (*tren*) [TOXIC]
Isophthalaldehyde
Methanol [FLAMMABLE]

Equipment
Conical flask (500 mL)
Hotplate/stirrer and stirrer bar
Pressure equalized dropping funnel
Condenser
Büchner funnel

Note: This reaction should be carried out in a fume hood

Add isophthalaldehyde (1.01 g, 7.5 mmol) to methanol (200 mL) in a 500 mL conical flask equipped with a stirrer bar. Stir to dissolve the solids. Add a solution of *tren* (0.73 g, 0.75 mL, 5.0 mmol) in methanol (20 mL) dropwise (*ca.* 15 min) to the clear solution which turns pale yellow during the addition. Once all the *tren* has been added reflux for 45 min then allow the solution to cool to room temperature. Transfer to a round-bottomed flask and reduce the volume to *ca.* 75 mL on the rotovap. Place the flask in an ice bath for 30 min then isolate the product, *tren*-derived cryptand (**37**), by filtration.

Yield: 0.55 g (~40%); m.p.: >250 °C; IR (v, cm^{-1}): 2830, 1645, 1355, 1290, 1035, 930, 695; ^{1}H NMR (δ, ppm; CDCl$_3$) 8.2 (2 × s, 9 H, ArH and N = CH), 7.6 (s, 9 H, ArH), 7.5 (t, 3 H, ArH), 3.8, 3.4, 2.9, 2.7 (br s, 24 H, CH_2).

Reduction of the tren-*cryptand (38)*

Reagents
Tren-Cryptand (**37**)
Ethanol (anhydrous) [FLAMMABLE]

Equipment
Round-bottomed flask (100 mL)
Heater/stirrer and stirrer bar

Potassium borohydride [CORROSIVE;   Glassware for extraction and work-up
   REACTS VIOLENTLY WITH WATER]
Ammonium chloride (2 M, aqueous solution)
Chloroform [TOXIC; CARCINOGEN]
Distilled water
Anhydrous magnesium sulphate

Note: This reaction should be carried out in a fume hood. Exercise due care when handling potassium borohydride.

Stir *tren*-cryptand, **37**, (0.73 g, 1.25 mmol) in methanol (100 mL) in a 100 mL round-bottomed flask equipped with a stirrer bar. Place in a heating mantle and bring to reflux. Cautiously add potassium borohydride (2.4 g, 45 mmol [sixfold excess]), in very small portions, down the reflux condenser. Wait until effervescence has subsided between additions. During this time the sparingly soluble cryptand dissolves as it is reduced. Once all the borohydride has been added (*ca.* 1 h), rinse any remaining solids down the condenser with a little ethanol (*ca.* 5 mL). Reflux for a further 2 h then remove from heat and stir until the reaction mixture reaches room temperature. Filter the solution to remove (and carefully dispose of) any unreacted borohydride, fit a calcium chloride guard tube and stir overnight at room temperature. Remove the solvent under vacuum and add a solution of ammonium chloride (20 mL, 2 M aqueous solution) to the crude product. Extract the resulting solution and any solids with chloroform (1 × 50 mL then 2 × 25 mL), wash with distilled water (30 mL) and dry the organic phase with magnesium sulphate (*ca.* 1 g). Filter and remove the solvent under vacuum to give the crude reduced *tren*-cryptand (**38**) as a colourless, hydroscopic oil.

Yield: 0.47 g (~60%); IR (v, cm^{-1}): 2830, 1340, 1290, 1060, 925, 695; ^{1}H NMR (δ, ppm; CDCl$_3$): 7.2 (m, 3 H, ArH), 7.1 (m, 9 H, ArH), 3.6 (m, 12 H, ArCH_2), 3.0–2.8 (m, 24 H, NCH_2CH_2N), 2.7 (br s, 6 H, NH).

4
Supramolecular Assembly

4.1 Detection, Measurement, Prediction and Visualization

Several methods may be employed to find out how molecules interact in supramolecular assemblies depending on the physical state of the system under observation (solid, solution, liquid or gas) and the purpose of the observation. The purpose of this chapter is to summarize a range of analytical techniques that can be applied to supramolecular assemblies in order to discover information about the forces holding them together, the complexity of the supramolecular systems themselves and, in the case of computational methods, to suggest possible mechanistic interpretations for analytical data through the application of theoretical models. The methods considered below are primarily concerned with the solid state (crystallography and solid state nuclear magnetic resonance (NMR) spectroscopy) and solution phase (NMR and optical spectroscopy). As most supramolecular phenomena are based on weak interactions there are few opportunities, other than mass spectrometry and gas chromatography, to use gas phase techniques. A commentary on solid state vs. solution behaviour is given to illustrate the strengths and weaknesses of both methods of analysis.

4.2 X-ray Crystallography

Supramolecular assemblies are, by definition, complex affairs and often generate spectral data that are hard to interpret. If the system can resolve into regular single crystals then the possibility of determining the solid-state structure evolves. From experience, supramolecular systems tend to fall into one of two classes in the crystalline state. They may produce wonderfully rigid one-, two- or three-dimensional

A Practical Guide to Supramolecular Chemistry Peter J. Cragg
© 2005 John Wiley & Sons, Ltd

lattices that give the crystallographer no problems and, due to the high degree of symmetry inherent in the structure, require relatively little data to be collected to yield a high quality structure. Because the lattice is so well preserved these crystal structures can often be solved from quite poor quality crystals although, given the system's drive to form a regular structure, poor quality crystals are rarely a problem under these circumstances. Good examples of these compounds are the molecular boxes of the Thomas group [1]. In this case the combination of transition metals with preferences for octahedral symmetry, capped on one face by [9]ane-S$_3$ and coordinated to C_2-symmetric ligands with rigid, linear, divergent bifunctional donor atoms leads to cubic supramolecules, as shown in Figure 4.1.

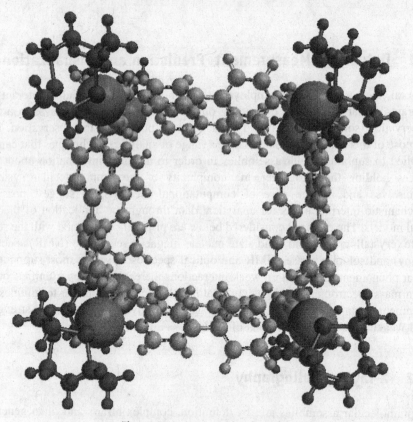

Figure 4.1 A supramolecular cube

Conversely, the sheer complexity of the supramolecular system may, if it crystallizes at all, lead to a very poorly defined structure. The first problem is then in growing a suitable crystal. If volatile solvents are in any way involved in the crystal lattice the chances are that they will evaporate before or during the data-gathering phase of the crystallographic experiment. Failing this the molecular assemblies may not occupy the same relative positions in adjacent crystal unit cells and thus generate disorder in the structure. In some cases it is possible to resolve this disorder by assuming models that require the molecules to adopt slightly different geometries every third or fourth unit cell. This allows the crystallographer to refine the atomic coordinates for some molecules differently to others and may result in 67:33 or 75:25 disordered systems. A good example of this is where a poorly resolved azacrown ether is involved. The electron density associated with oxygen (six valence electrons) is the same as that of an amine (five valence nitrogen electrons and one from hydrogen). If the crown ethers in neighbouring unit cells change relative orientation, such that the position occupied by nitrogen in one is occupied by oxygen in the next, it will appear from the crystallographic data that the atom has characteristics which are 50 per cent oxygen and 50 per cent nitrogen. This in turn can lead to problems in interpreting supramolecular phenomena, for example, the importance of hydrogen bonding in the structure. Fortunately the recent advances in X-ray diffractometers and crystallographic software have largely alleviated this particular problem.

A greater challenge that comes with increasing supramolecular complexity, particularly with regard to capsular systems, is that of the refinement of guest orientation. On occasion it is possible to determine the presence of solvent molecules but not their positions. An excellent example of this is in the supramolecular capsule described by MacGillivray and Atwood [2]. In the structure six [4]resorcinarenes and eight water molecules self-assemble to form a capsule that has the shape of a snub cube, one of the 13 Archimedean solids. The cavity is of the order of 1400Å^3 (1.4nm^3) and contains at least 20 solvent molecules, though none of them can be identified crystallographically. In a later example the crystals of the analogous pyrogallol[4]arene grown from ethyl acetate solution allowed not only the identification of all the included solvent molecules but also their orientations [3].

The main problem in growing suitable crystals and having them diffract to a crystallographer's satisfaction is usually in the nature of the supramolecular components themselves. In the absence of a heavy atom, such as a metal or main group element with a high atomic mass, 'direct methods' of solving crystal structures become necessary. For large organic molecules, and supramolecular complexes, this approach to solving crystal structures can be problematic. It can lead to the disorder problem described above that makes the different binding sites in mixed oxa- and azacrowns hard to distinguish; were these sites to interact with a transition metal the resolution would be easier. A second problem exists when the molecules are conformationally mobile. Although it is routine to collect X-ray data at liquid nitrogen temperatures, thus minimizing atomic thermal motion, subtle differences in conformers may overcomplicate the structure solution. An example of this is the complex formed between 4-t-butyloxacalix[3]arenetris(*N,N*-diethyl) acetamide (**31**) and mercury(II) chloride [4]. One mercury atom binds in a bidentate fashion with a phenolic oxygen, an amide oxygen and two chlorides,

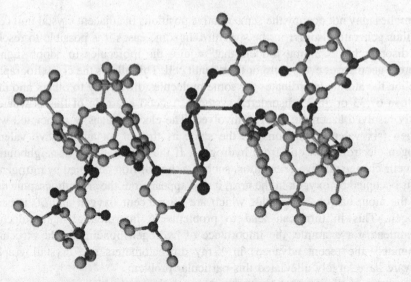

Figure 4.2   A dimeric complex with apparent C_2 symmetry

one of which bridges to another mercury atom, to give a symmetric Hg_2Cl_4 core. This would suggest C_2 symmetry but unfortunately, although the central motif does have this relationship, the alkyl tails of the macrocycles do not follow it perfectly, nor are they simply disordered. The final solution, as shown in Figure 4.2, had to assume a much less symmetric supramolecule and required a large data collection. Despite the apparent simplicity of the symmetry involved in the supramolecular nature of the complex, the conformational mobility of the ligands involved meant that the X-ray structure was much harder to resolve than initially anticipated.

Surely, given the wealth of crystallographic evidence for supramolecular systems, there must be protocols to follow that will produce good quality crystals from these complex assemblies? Unfortunately there is no guaranteed method of growing crystals, supramolecular or otherwise, suitable for single crystal X-ray diffraction methods. In the case of supramolecular assemblies this is an even greater problem as the individual molecules often interact through hydrogen bonds or even weaker effects. Added to this the ligands often have a high degree of conformational flexibility, which makes them even less amenable to form regular crystalline structures. Fortunately, the supramolecular motif is often more important than the exact positions of all the species in the assembly which allows the publication of crystal structures with poorer levels of refinement than are usually acceptable for scholarly journals. A similar case can be made in protein crystallography where the resolution may be good enough to determine the sequence of amino acids, and the overall structure of the molecule, but not necessarily the exact orientation of the side chains. The latter information is important, and may be the focus of later investigations, but initially it is the gross structural features that are of interest.

Occasionally, a complex does spontaneously form good crystals from the reaction mixture. Failing this the product should be recrystallized from a solvent, or mixture of solvents, in which it is ideally insoluble at room temperature and infinitely soluble at reflux. If this simple expedient does not work then a solvent must be found in which the complex is highly soluble. Cooling a concentrated solution over an extended period of time may deposit crystals. Alternatively, slow evaporation of the solvent from a dilute solution, either at room temperature or below, may be more successful. Another approach is to carefully add a layer of a less dense solvent on top of a dilute solution containing the complex and hope to observe crystal growth at the interface. The two solvents should be immiscible or nearly so. One final method is to leave a solution in an open topped vessel in an enclosed environment, such as a dessicator, that contains either a desiccant or a vessel containing a second solvent in which the complex is insoluble. Over time the atmosphere in the closed system will equilibrate leading to small changes in the composition of the solution and complex crystallization (Figure 4.3). Note that changes in solvents will influence the supramolecular structure that finally appears. Adding hexane to diethyl ether, or dichloromethane to ethanol, will

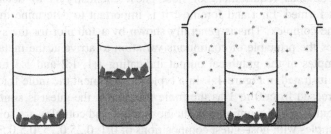

Figure 4.3 Strategies for crystallization: directly from solvent (left), use of a secondary solvent (centre) and in a controlled atmosphere using a secondary solvent (right)

undoubtedly disrupt any hydrogen bonding that may have held the complex together. It is also worth bearing in mind that solvents with low boiling points easily escape from crystal lattices. Thus, the perfect crystals that grew in the freezer may crumble to an amorphous powder at room temperature under the X-ray beam. Finding the correct solvent or solvent mixture from which to grow good quality crystals is usually a matter of trial and error. For example, acetonitrile and acetonitrile – methanol mixtures appear to be particularly successful media from which to grow good crystals of polyether and crown ether complexes of lanthanide metals [5,6]. Perhaps the only certainty regarding crystal growth is that whatever conditions work for one particular complex will probably work for similar systems!

[1] Self-assembly of a supramolecular cube, J. A. Thomas, S. Roche, C. Haslam, H. Adams and S. L. Heath *Chem. Commun.*, 1998, 1681.

[2] A chiral spherical molecular assembly held together by 60 hydrogen bonds, L. R. MacGillivray and J. L. Atwood, *Nature*, 1997, **389**, 469.

[3] Inner core structure responds to communication between nanocapsule walls, G. W. V. Cave, J. Antesberger, L. J. Barbour, R. M. McKinlay and J. L. Atwood, *Angew. Chem. Int. Ed.*, 2004, **43**, 5263.

[4] Implications for mercury toxicity from the structure of an oxacalix[3]arene-HgCl$_2$ complex? P. J. Cragg, M. Miah and J. W. Steed, *Supramol. Chem.*, 2002, **14**, 75.

[5] Macrocycle complexation chemistry 35. Survey of the complexation of the open-chain 15-crown-5 analogue tetraethylene glycol with the lanthanide chlorides, R. D. Rogers, R. D. Etzenhouser, J. S. Murdoch and E. Reyes, *Inorg. Chem.*, 1991, **30**, 1445.

[6] Structural investigation into the steric control of polyether complexation in the lanthanide series – macrocyclic 18-crown-6 versus acyclic pentaethylene glycol, R. D. Rogers, A. N. Rollins, R. D. Etzenhouser, E. J. Voss and C. B. Bauer, *Inorg. Chem.*, 1993, **32**, 3451.

4.3 Spectroscopic and Spectrometric Techniques

The most obvious requirement of host–guest chemistry is to determine if a complex is formed. First and foremost it is important to determine the stoichiometry of the complex. This is generally shown by a Job plot for the system. The method uses the principle of continuous variation to arrive at the metal to ligand ratio. Examples of the graphical output indicating 1:1, 1:2 and 1:3 metal:ligand ratios are illustrated in Figure 4.4. In a typical experiment the mole fraction of the host is increased from 0 to 1 as the mole fraction of the guest is simultaneously reduced from 1 to 0. Assuming that the host, guest and complex are soluble in the solvent, samples with host–guest compositions of 0:1, 0.25:0.75, 0.5:0.5, 0.75:0.25 and 1:0 can be prepared and allowed to come to equilibrium. Next it is important to have a feature that is unambiguously indicative of the complex. This may be a new

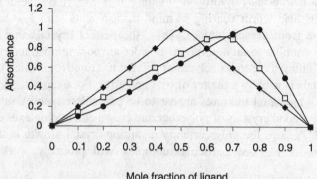

Figure 4.4 Typical Job plots for 1:1 metal:ligand (◆), 1:2 metal:ligand (□) and 1:3 metal:ligand (●) complex formation

feature in the optical spectrum, a set of shifted peaks in the NMR spectrum or a change in the redox potential. In the case of very stable complexes, the formation of new mass peaks in the mass spectrum could be used for this purpose. The relative intensities of the complex-derived feature are plotted against the composition of each sample to obtain the stoichiometry. A 1:1 complex is indicated by a maximum intensity for the mixture with a composition of 0.5:0.5 host:guest ratio. A maximum value for a 0.25:0.75 host:guest mixture would indicate that three guests were bound by each host molecule.

A second desirable measurement might be extractability of different guests by a specific host in order to determine its specificity for a particular guest in the presence of other potential interfering species. For example a macrocycle may have been designed to extract radioactive caesium from dilute acid, contaminated by low-grade nuclear waste, into an organic phase. In the laboratory it would be prudent to establish how well the macrocycle extracted all the other alkali metals and possibly the alkaline earths too. Further experiments may involve mixtures of cations and a range of acidities to model, as closely as possible, the actual operating conditions. For this experiment it is important that a parameter can be established that allows the extracted species to be monitored. It would be advisable to test the efficacy of the macrocycle by extracting metal picrate salts from an acidic solution into an organic phase containing the macrocycle. This method allows the extraction to be followed by colorimetric methods as the picrate anion is intensely yellow and can be detected at 390 nm even at low concentrations. With luck the macrocycle itself will contain a chromophore with a well-defined absorbance maximum that shifts upon binding a guest. In this case it may be possible to detect differences between guests and determine specificity in the presence of interfering species.

Finally, it is often useful to determine binding constants, K_a, for supramolecular complexes. K_a incorporates the effects of complex association and dissociation through the relationship:

$$K_a = K_1/K_{-1} \quad (\text{or} \log K = \log K_1 - \log K_{-1})$$

where K_1 = association constant and K_{-1} = dissociation constant.

The overall binding constant for a simple host–guest system such as:

$$[H] + [G] \leftrightarrow [H \cdot G]$$

where [H], [G] and [H·G] are the concentrations of each species at thermodynamic equilibrium, is given by:

$$K_a = [H \cdot G]/[H][G]$$

and is therefore is a measure of the entire system at thermodynamic equilibrium. It can be related directly to the free energy of complexation by the Gibbs equation:

$$\Delta G^\circ = -RT \ln K_a$$

An excellent, and far more thorough, treatment of binding constants can be found in Connors' book of the same name [1] and an exhaustive treatment of supramolecular complexation thermodynamics has been undertaken by Schneider and Yatsimirsky [2]. A highly relevant review comparing methods for determining supramolecular complex stabilities was also published in 1992 [3].

[1] *Binding constants*, K. A. Connors, John Wiley & Sons, Ltd, Chichester, 1987.
[2] *Principles and Methods in Supramolecular Chemistry*, H.-J. Schneider and A. Yatsimirsky, John Wiley & Sons, Ltd, Chichester, 2000.
[3] A comparison of different experimental techniques for the determination of the solubilities of polyether, crown ether and cryptand complexes in solution, H.-J. Buschmann, *Inorg. Chim. Acta*, 1992, **195**, 51.

4.4 Binding Constant Determination

Of the many methods of determining binding constants the two most commonly encountered in the context of supramolecular chemistry are based on ultraviolet/ visible spectroscopy and nuclear magnetic resonance spectroscopy. The former assumes that a new absorption peak forms upon host–guest complexation thus concentrations of host, guest and complex can be calculated for a variety of stoichiometries. This method, based on the oft-quoted paper of Benesi and Hildebrand [1] is well suited to complexes in which either host or guest have chromophores. The latter method is more widely applied as many supramolecular hosts do not contain chromophores; however, their protons are subject to measurable shifts upon guest binding. The guests may also exhibit shifts that can be used to determine binding constants. Secondary information can also be gleaned by following host-guest binding by ^{1}H NMR as those protons that interact most closely during complexation will also have the greatest shifts when compared to spectra of the uncomplexed species. A wide-ranging review of NMR approaches to association constant determination was published in 2000 [2].

Both of the above methods require that the stoichiometry is known prior to manipulation of the data. Most experimentalists assume 1:1 stoichiometry, though this is not always the case, and that self-association of the host does not occur. These two possibilities may be checked, in the first case, by constructing a Job plot and, in the second, by undertaking dilution experiments of the host alone. It may also be pertinent to attempt spectroscopic studies in a solvent where self-association may be enhanced and another that would inhibit the effect. For example an organic host bearing a carboxylic acid group may be expected to dimerize in an aprotic solvent such as dichloromethane or deuterochloroform but not in methanol or its deuterated analogue. Assuming that the host was soluble in both solvents it would be easy to determine if self-association through carboxylic acid dimerization was occurring.

For cases of 1:1 stoichiometry the following relationships apply:

$$[H] + [G] \leftrightarrow [H \cdot G] \tag{1}$$

$$K_a = [H \cdot G]/[H][G] \tag{2}$$

$$[H]_0 = [H] + [H \cdot G] \tag{3}$$

$$[G]_0 = [G] + [H \cdot G] \tag{4}$$

where K_a = binding constant, $[H]_0$ = initial concentration of host, $[G]_0$ = initial concentration of guest, $[H]$ = concentration of host at equilibrium ('uncomplexed host'), $[G]$ = concentration of guest at equilibrium ('uncomplexed guest') and $[H \cdot G]$ = concentration of host–guest complex at equilibrium.

By analogy to the Benesi–Hildebrand method for determining K_a using ultraviolet/visible spectroscopic absorbances, several groups have derived NMR-based determinations of K_a. It is assumed that upon formation of the complex species $H \cdot G$ chemical signals from protons in the one or both of the species will shift with respect to those in the uncomplexed host and guest. Depending on whether the equilibrium set up in (1) is fast or slow one of two approaches can be made by NMR to determine the binding constant. The definitions of fast and slow relate to the rate constant of the binding event and its relationship to the chemical shift. In the case of fast exchange only one signal, a population average, is seen for the complexed and uncomplexed protons in question. Slow exchange gives rise to two signals, one in the uncomplexed and one for the complexed protons, that disappear and appear in a reciprocal manner. Fast exchange implies that the shift (in Hz) is larger than the rate constant, in slow exchange the shift is smaller than the rate constant.

In the case of slow exchange the binding constant can be determined using equation (2) directly. As both the bound and free guests generate signals their relative concentrations can be calculated from the integration of the sets of peaks arising from the two states. The value of $[H \cdot G]$ will come from the new peaks that appear on the spectrum. The values of $[H]$ and $[G]$ can be obtained from the integration of the related peaks coming from uncomplexed species although it is more appropriate to relate these values to those of the original solutions, $[H]_0$ and $[G]_0$. To relate concentrations of species in solution it is necessary to combine equations (2), (3) and (4) to give:

$$K_a = [H \cdot G]/([H]_0 - [H \cdot G])([G]_0 - [H \cdot G]) \tag{5}$$

For slow exchange systems this allows the original concentrations to be used to calibrate the system. Combining the integrals of sets of peaks generated by specific protons in bound and free guests gives a total integral equivalent to $[G]_0$. A similar exercise can be done to correlate host integration with $[H]_0$. The complex concentration, $[H \cdot G]$, can then be calculated and, as $[H]_0$ and $[G]_0$ are known, K_a determined using equation (5).

For fast exchange the calculation is more complicated and requires that averaged signals for protons that are affected by complexation are distinct from the signals those same protons generate in the unbound state. It is also desirable to reduce the number of variables in the equation. This can be achieved by using dilute solutions of either host or guest. In dilute solutions where $[G]_0 \gg [H]_0$ then:

$$[G]_0 - [H \cdot G] \approx [G]_0 \qquad (6)$$

Or alternatively, if $[H]_0 \gg [G]_0$ then:

$$[H]_0 - [H \cdot G] \approx [H]_0 \qquad (7)$$

These relationships allow the determination of the binding constant by monitoring the chemical shift of protons on either the host or guest. Ideally high field instruments (500 or 600 MHz) should be used for their high resolution although good quality data can still be obtained on lower field instruments. Proton NMR is far more informative for this purpose than carbon-13 as the nuclei of the latter rarely shift far upon complexation. Proton shifts, on the other hand, may be of the order of 1 to 2 ppm and easy to detect even on low field instruments.

Substituting for $[G]_0 - [H \cdot G]$ in equation (6) above, in which a large excess of guest is present, the following is obtained:

$$K_a = [H \cdot G]/([H]_0 - [H \cdot G])([G]_0) \qquad (8)$$

Defining the maximum difference in chemical shifts for a specific guest proton in the uncomplexed state and the host–guest complex as $\Delta\delta_{max}$, and the difference between the uncomplexed shift of the same guest proton, as the host is added, as $\Delta\delta$ the following is found:

$$[H \cdot G]/[G]_0 = \Delta\delta/\Delta\delta_{max} \qquad (9)$$

Substitution into equation (8) and rearrangement gives:

$$\Delta\delta/[H]_0 = -K_a \Delta\delta + K_a \Delta\delta_{max} \qquad (10)$$

A linear plot of $\Delta\delta/[H]_0$ against $\Delta\delta$ will have a slope of $-K_a$ with an intercept of $\Delta\delta_{max}$. This particular use of NMR data is derived from the work of Foster and Fyfe [3], which in turn is a variation on the well-known Scatchard plot [4]. It differs slightly from the Hanna–Ashbaugh approach [5], itself a method of applying the Benesi–Hildebrand solution to NMR, that is also widely used. For the data to be valid the chemical shifts of the guest protons must relate to the species in equilibrium, thus a large excess of the host has to be present: most literature examples use a tenfold excess or higher. The experiment therefore requires a series of solutions with a range of initial host concentrations, with the host always in excess, assuming that equation (9) is to be used. These concentrations give known values for $[H]_0$. For each solution $\Delta\delta$ values are recorded for guest protons, ideally using those signals that give the greatest shift over the concentration range to give the highest resolution and accuracy. Plotting $\Delta\delta/[H]_0$ against $\Delta\delta$ gives the binding constant, K_a. An example of this method is given here. It illustrates the interactions between the oxacalix[3]arene and quaternary ammonium cation shown in Figure 4.5.

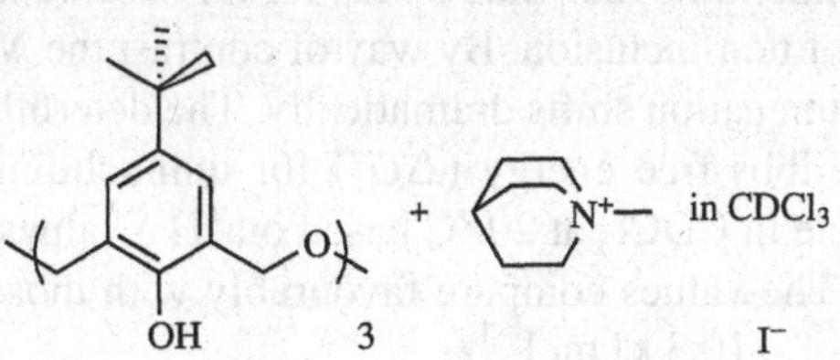

Figure 4.5 Quinuclidinium binding by an oxacalix[3]arene

4-t-Butyloxacalix[3]arene (**28**) has a ^1H NMR spectrum in $CDCl_3$ consisting of singlets at 8.6 (3 H, OH), 6.9 (6 H, ArH), 4.7 (12 H, OCH_2Ar) and 1.3 ppm (27 H, t-butyl CH_3). The relative simplicity of the spectrum makes any guest binding relatively easy to observe as there are many useful 'windows' between signals where guest signals may be clearly detected without overlap with ligand peaks. Given the threefold symmetry of the oxacalixarene it might be expected to bind guests with complementary symmetry. Two potential guests, quinuclidine and N-methylquinuclidinium iodide, illustrate the importance of cation–π interactions in guest inclusion. The latter is readily prepared in a fume hood by adding a solution of methyl iodide (CAUTION: NEUROTOXIN) in dry diethyl ether (CAUTION: FLAMMABLE) to a stoichiometric amount of quinuclidine, also in dry diethyl ether. The quaternary ammonium salt forms as a white precipitate and can be isolated by filtration.

Determining binding constants by NMR

Binding constant determination by ^1H NMR: oxacalix[3]arene inclusion complexes

Reagents
4-t-Butyloxacalix[3]arene (**28**)
N-Methylquinuclidinium iodide
Deuterochloroform ($CDCl_3$)

Equipment
Volumetric flasks (5 mL)
Syringes/pipettes for accurate dilution
NMR tubes

Make up solutions of the host and guests in $CDCl_3$ at 100 mM concentrations in 5 mL volumetric flasks. Record the ^1H NMR spectra of all three compounds. Next use the stock solutions to prepare two solutions that are 90 mM with respect to the oxacalixarene and 10 mM with respect to the guest. Record the spectra of both. Prepare a series of host–guest solutions to give progressively dilute samples down to a host concentration of 1 mM. Record the spectra of all the diluted solutions. Tabulate the chemical shift (δ) for each guest proton environment against host concentration ($[H]_0$). Choose the signal that gives the greatest shift over the course of the experiment and calculate $\Delta\delta$ for that signal. Finally plot $\Delta\delta/[H]_0$ against $\Delta\delta$ to determine K_a for both complexes.

In the case of quinuclidine the value of $\Delta\delta$ for all observable protons is so small that it is indicative of non-inclusion. By way of contrast the N-methyl group of the quaternary ammonium cation shifts dramatically. The determination of the binding constant (K_a) and Gibbs free energy ($\Delta G°$) for quinuclidinium iodide with 4-t-butyloxacalix[3]arene in $CDCl_3$ at 20 °C based on ^1H δ values for NCH_3 protons is given in Table 4.1. The values compare favourably with those in the literature [6] which give a $\Delta G°$ of $-10.3\,kJ\,mol^{-1}$.

Table 4.1 Binding constant determination for a quinuclidinium–oxacalixarene complex

$[H]_0$ (M)	δ	$\Delta\delta$	$\Delta\delta/[H]_0$
0	3.757	–	–
3.00×10^{-2}	2.478	1.279	42.63
1.50×10^{-2}	2.644	1.113	74.20
7.50×10^{-3}	2.836	0.921	122.80
3.75×10^{-3}	3.036	0.721	192.27

$K_a = 268\,M^{-1}$; $\Delta G° = -13.6\,kJ\,mol^{-1}$

[1] A spectrophotometric investigation of the interaction of iodine with aromatic hydrocarbons, H. A. Benesi and J. H. Hildebrand, *J. Am. Chem. Soc.*, 1949, **71**, 2703.

[2] Tetrahedron Report Number 536. Determination of association constants (K_a) from solution NMR data, L. Fielding, *Tetrahedron*, 2000, **56**, 6151.

[3] Interaction of electron acceptors with bases. Part 15. Determination of association constants of organic charge-transfer complexes by n.m.r. spectroscopy, R. Foster and C. A. Fyfe, *Trans. Faraday Soc.*, 1965, **61**, 1626.

[4] The attractions of proteins for small molecules and ions, G. Scatchard, *Ann. N. Y. Acad. Sci.*, 1949, **51**, 660.

[5] Nuclear magnetic resonance study of molecular complexes of 7,7,8,8-tetracyanoqui-nodimethane and aromatic donors, M. W. Hanna and A. L. Ashbaugh, *J. Phys. Chem.*, 1964, **68**, 811.

[6] Homooxacalixarenes 3. Complexation of quaternary ammonium ions by parent homo-oxacalixarenes in $CDCl_3$ solution, B. Masci, *Tetrahedron*, 1995, **51**, 5459.

4.5   Solid State vs. Solution Behaviour

A common statement made about crystal structures is that they give a true depiction of the compound of interest as the technique is able to determine the relative positions of atoms within the structure exactly. Even if this were true it only gives information for one crystalline form of the compound that is, by necessity, in a state of lower solvation than the solution phase. The technique also gives no information on the dynamic behaviour of the compound or supramolecular complex. In fact the method by which X-ray structures are determined relies heavily on the expertise of the crystallographer and the software used to interpret the experimental data. On the other hand, diffraction data from single crystals does give extremely useful information about supramolecular phenomena, such as the influence of solvent polarity or hydrogen-bonding potential on intermolecular interactions, or the relative positions of host and guest molecules. Of particular relevance to supramolecular chemists is the ability to determine the existence of hydrogen bonds between or within molecules. While the accurate detection of hydrogen atoms in X-ray crystallography is sometimes difficult, it *is* possible to look for hydrogen-bonded atoms specifically as the bond lengths and angles of likely candidates are generally distorted from the norm and thus they make their presence felt to an experienced crystallographer.

Information about relative positions of functional groups or the size of cavities within macrocyclic systems is clearly of use when analysing supramolecular systems or designing new macrocycle derivatives and, for these reasons, crystal structures are a vital part of the supramolecular chemist's library. Indeed, much of our early knowledge of supramolecular phenomena came about through the use of X-ray crystallography giving, as it did, information about the optimum cation size for crown ether complexation or the unexpected layer-like structure of the sodium sulphonatocalix[4]arene salts, to pick just two examples.

4.6   Supramolecular Chemistry *In Silico*: Molecular Modelling and Associated Techniques

Molecular modelling is perhaps one of the most useful techniques available to chemists interested in designing supramolecular synthons or modifying their properties. However, despite the astounding advances in computational power and improvements in software over the past few decades, it must be stated first and foremost that results of computer-generated simulations are no substitute for laboratory-based experimentation. It is worth noting the primary dictionary definition of 'simulation' is 'to assume the outward qualities or appearance of (something), usually with the intent to deceive' [1]! Computational approaches can be used to simulate atomic, molecular and supramolecular behaviour thereby generating information about any number of properties. The methods used may be based on fundamental descriptions of atomic and molecular orbitals (*ab initio* quantum mechanics), experimental data (*a priori* molecular mechanics) or a combination of both (semiempirical methods). The choice of method depends on the task in hand and the computational resources available. The most computationally intensive methods are based on calculations of molecular orbitals from solutions to the Schrödinger equation. These *ab initio* methods use fundamental mathematical principles to derive the structures and properties associated with molecular orbitals through a consideration of the effect of every electron in the molecule. The number of calculations, and their complexity, limits the size of molecules which may be studied by these means using conventional desktop machines.

'Computer-aided molecular design', or *in silico* chemistry, is of great value in visualizing supramolecular systems and in probing the likely effects of structural modification. For example, which will form the stronger complex with a sodium cation, [15]crown-5 or [18]crown-6? What about the analogous potassium complex? A literature search or laboratory experiment designed to answer this question would require considerable time and effort. The simulation may take only a few minutes. A similar question may be raised regarding the potential for self-complementary species to form supramolecular complexes. These two examples will be used later to illustrate the application of computational methods to problems in supramolecular chemistry.

Although computational methods have been available for several decades they have only been applied to problems of a supramolecular nature relatively recently. Most of the early work was on relatively simple systems and employed molecular mechanics [2–5]. Often these required extensive parameterization to accurately represent host-guest interactions. With advances in computational power it has been possible to model molecular dynamics [6,7], solvent interface behaviour [8], guest binding [9], free energy calculations for host–guest systems [10] and even high-level quantum mechanical approaches to predict binding mechanisms [11] and NMR shifts [12]. As a testament to the many applications of computational chemistry to the supramolecular field an entire volume of the NATO ASI series was devoted to the intersection between the two subjects as early as 1994 [13].

In silico supramolecular chemistry can be used in conjunction with other forms of analysis. One valuable source of data, the Cambridge Structural Database, now contains so many X-ray structures that it is increasingly being mined for information about host–guest systems. In particular, it can be used to compile examples of intermolecular hydrogen bonding that can in turn be used to inform the design of host molecules required for specific guests. One of the most recent examples where *in silico* predictions have been combined with crystallographic data is Hay's determination of ideal hydrogen bonding environments necessary for the successful design of receptors for the nitrate anion [14].

[1] *The New Penguin English Dictionary*, Penguin Books, London, 2001, p. 1305.

[2] A molecular mechanics study of 18-crown-6 and its alkali complexes – an analysis of structural flexibility, ligand specificity, and the macrocyclic effect, G. Wipff, P. Weiner and P. Kollman, *J. Am. Chem. Soc.*, 1982, **104**, 3249.

[3] Structural criteria for the rational design of selective ligands: extension of the MM3 force field to aliphatic ether complexes of the alkali and alkaline earth cations, B. P. Hay and J. D. Rustad, *J. Am. Chem. Soc.*, 1994, **116**, 6316.

[4] Combined NMR-spectroscopy and molecular mechanics studies on the stable structures of calix[n]arenes, T. Harada and S. Shinkai, *J. Chem. Soc., Perkin Trans. 2*, 1995, 2231.

[5] Molecular mechanics studies of a series of dicopper complexes of a 20-membered macrocycle containing differing types of bridges between the metal centers, M. G. B. Drew and P. C. Yates, *J. Chem. Soc., Dalton Trans.*, 1987, 2563.

[6] Macrocyclic thioether design by molecular modeling, G. A. Forsyth and J. C. Lockhart, *Supramol. Chem.*, 1994, **4**, 17.

[7] Molecular dynamics simulations of p-sulfonatocalix[4]arene complexes with inorganic and organic cations in water: A structural and thermodynamic study, A. Mendes, C. Bonal, N. Morel-Desrosiers, J. P. Morel and P. Malfreyt, *J. Phys. Chem. B*, 2002, **106**, 4516.

[8] Distribution of hydrophobic ions and their counterions at an aqueous liquid–liquid interface: A molecular dynamics investigation, B. Schnell, R. Schurhammer and G. Wipff, *J. Phys. Chem. B*, 2004, **108**, 2285.

[9] An *ab initio* study of the influence of crystal packing on the host–guest interactions of calix[4]arene crystal structures, M. I. Ogden, A. L. Rohl and J. D. Gale, *Chem. Commun.*, 2001, 1626.

[10] Computational approaches to molecular recognition, M. L. Lamb and W. L. Jorgensen, *Curr. Opin. Chem. Biol.*, 1997, **1**, 449.

[11] Halide anion recognition by calix[4]pyrrole: a quantum chemical study, F. Pichierri, *J. Mol. Struct. (Theochem)*, 2002, **581**, 117.

[12] GIAO-DFT calculated and experimentally derived complexation-induced chemical shifts of calix[4]arene-solvent inclusion complexes, A. C. Backes, J. Schatz and H. U. Siehl, *J. Chem. Soc., Perkin Trans. 2*, 2002, 484.

[13] *Computational Approaches in Supramolecular Chemistry*, NATO ASI Series vol. 426, G. Wipff, Ed., Kluwer Academic Publishers, Dordrecht, 1994.

[14] Structural criteria for the rational design of selective ligands: convergent hydrogen bonding sites for the nitrate anion, B. P. Hay, M. Gutowski, D. A. Dixon, J. Garza, R. Vargas and B. A. Moyer, *J. Am. Chem. Soc.*, 2004, **126**, 7925.

4.7 Computational Approaches

Supramolecular chemistry is primarily a science in which host molecules are designed to perform a particular function upon forming a complex with a guest. It is therefore of paramount importance that any computational treatment is able to generate a model that contains accurate information about the geometry of the host or host–guest complex. In this we are looking for similar insights to those generated by X-ray crystallography or two-dimensional NMR experiments. How do the host's size, shape, electrostatic field and conformational preferences affect the way that it binds guests? Why are particular complexes more stable than others? How could particular properties, such as aqueous solubility or luminescence, be enhanced without compromising the strength and specificity of guest inclusion by the parent compound? Questions such as these can be answered by very experienced supramolecular chemists, but for the rest of us a computer model often gives the insights that would otherwise only come with extensive searches of the literature. The potential benefits and limitations of computational methods must be understood if this powerful technique is to be used as another tool in the service of supramolecular chemistry.

Molecular mechanics methods

Molecular mechanics methods represent the simplest computational approaches and may be found in a large array of computational software suites. The general technique assumes that atoms are hard spheres and are attached to each other by elastic bonds that obey Hooke's Law. Atoms interact at an optimum distance that is dictated by the 'strength' of the elastic bond (the force constant) and an 'ideal' bond length. The latter is usually based on an average of many similar bonds determined by a variety of techniques. For example the ideal carbon–carbon bond length in an alkane is 1.41 Å (0.141 nm) as determined for thousands of examples using a variety of methods. A similar method is used to determine bond angles and torsions. As well as these effects of bonded atoms it is important to include electrostatic interactions using Coulombic, Lennard-Jones or similar relationships and other forces such as 'non-van der Waals 1–4 interactions'. For the software to generate a reasonable structure, parameters for all the terms must be recorded within a reference file usually called a force field. Different force fields are used depending on the application. For example, the parameters used by structural biologists to accurately describe amide interactions in proteins are unlikely to be of interest to inorganic chemists who want to know if a square planar or tetrahedral geometry is the more stable for a particular coordination complex. The aim of molecular mechanics methods is to generate a three-dimensional model of a compound in which the geometry is optimized to reduce overall strain in the system. Naturally, there are several ways of achieving this though they all involve moving part of the molecule and determining if the new geometry has a higher or lower energy. As long as the geometry resulting from the 'next move' is of a lower energy, and the atomic interactions are not too distorted from their ideal values, that move will be accepted. When certain criteria are met for sequential moves, such as energy differences between subsequent moves or the root mean square of the difference in atomic coordinates for the whole molecule being less than a specified value, the simulation will have been 'minimized' and a final steric energy calculated. The overall energy can often be broken down into various subcategories (bond stretch, van der Waals, and the like) so that the major influences on the molecule can be determined. This may be of use where factors like π–π stacking can be identified as a specific factor in the energy analysis and explains why the simulation results in the formation of dimeric supramolecules rather than discrete monomers.

Molecular mechanics are the most broadly applicable computational techniques as they can be used on small molecules as successfully as on proteins. The speed with which large numbers of simple calculations can be undertaken on even modest computers, together with the ease with which the method can be understood by non-specialists, engenders the broadest use of this technique.

As this method is the most commonly encountered, and is often the basis for geometries used in higher-level calculations, it is worth describing the interactions in detail. The molecular mechanics method assumes that the total energy of the system may be broken down into the following components:

$$E_{\text{total}} = \Sigma E_{\text{bond stretch}} + \Sigma E_{\text{bond angle bend}} + \Sigma E_{\text{torsional twist}} + \Sigma E_{\text{non-bonded terms}}$$

Thus by calculating and summing all the individual steric energy components involved in the optimized structure (bond stretch, bond angle bend, torsional or dihedral twist and non-bonded terms) an overall minimized energy can be determined. The different types of interactions are illustrated in Figure 4.6.

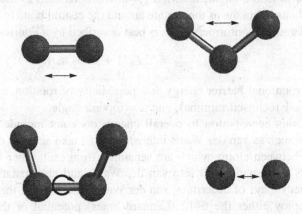

Figure 4.6 Molecular mechanics components: bond stretch (top left), bond angle bend (top right), torsional twist (bottom left), non-bonded interaction (bottom right)

The best description of a bond stretch is a Morse function but, as this is computationally expensive, a simpler harmonic function is usually used in molecular modelling. Many force fields use extra terms in the equation to improve the accuracy of the function which, at its simplest, has the form:

$$E_{\text{bond stretch}} = \Sigma k_1 (l - l_0)^2$$

where $k_1 =$ stretching force constant, $l =$ bond length and $l_0 =$ reference bond length.

The contribution which the bond energy makes to the overall structure is found by summing the energy for all bonds. Ideal values (l_0) are required for all types of bonds, thus C—C, C=C and C≡C bonds will need a different set of parameters to describe their behaviour.

Angles are treated in a similar way to bonds by using a harmonic function based on:

$$E_{\text{bond angle bend}} = \Sigma \, k_\theta (\theta - \theta_0)^2$$

where $k_\theta =$ angular force constant, $\theta =$ angle and $\theta_0 =$ equilibrium angle.

As with bond stretches, a different set of parameters must be defined for each type of atom to accurately model its behaviour. For example, an aromatic carbon in benzene will have a different equilibrium angle, θ_0, to a carbon in a ketone or aldehyde despite the apparent similarities in their sp^2 geometries. This approach can lead to problems particularly for square planar and octahedral coordination environments around transition metals. In these cases the angles between ligands need to be multiples of 90°. Unfortunately many early force fields often instructed angles less than 90° to optimize to 0°. The frustration that this caused, with multiple bonds collapsing into one, can be imagined.

Torsion, or dihedral, angles describe 1,4-interactions such as those between the terminal carbons in butane as they rotate around the central bond to give staggered, eclipsed and gauche conformers. This is best described by a Fourier series such as:

$$E_{\text{torsional twist}} = \Sigma \, E_n (1 + s \cos n\omega)$$

where $E_n =$ rotational barrier energy, $n =$ periodicity of rotation, $s = 1$ (staggered minima) or -1 (eclipsed minima), and $\omega =$ torsion angle.

Note that this contribution to overall energy does not include other through-space effects such as van der Waals interactions. To take account of these effects, interactions between atoms which are separated from each other by greater than 1,4 distances are usually split into van der Waals and electrostatic components. There are many ways of describing van der Waals interactions: the most common methods employ either the 6–12 (Lennard-Jones) potential or the Buckingham potential as shown below:

$$E_{\text{van der Waals}} = \Sigma \, \varepsilon [(r_m/r)^{12} - 2(r_m/r)^6] \qquad \text{6–12 potential}$$

where $\varepsilon =$ well depth and $r_m =$ minimum energy interaction distance.

$$E_{\text{van der Waals}} = \Sigma A \, \exp(-Br) - Cr^{-6} \qquad \text{Buckingham potential}$$

where A, B and C are constants.

The electrostatic term is usually calculated using partial charges by applying Coulomb's law:

$$E_{\text{electrostatic}} = \Sigma\, q_i q_j / D r_{ij}$$

where q_i, q_j = partial charges on atoms i and j, D = dielectric constant and r_{ij} = distance between i and j.

Other terms are often added to account for phenomena such as hydrogen bonds, so-called 'cross terms' and inversion of atomic centres through repositioning of lone pairs. It is this aspect of parameterization which is often seen as the method's failing: to get an accurate answer it is necessary to provide the program with accurate *a priori* data specific to the system under investigation. Fortunately in contemporary force fields extensive use is made of training sets of data to develop different atom and bond types. This, coupled to intelligent software that can interrogate structures to determine bond order and formal charge, has made many commercial molecular mechanics systems highly accurate.

In a molecular mechanics simulation the first stage is to generate a model. Most computational chemistry software now includes a graphical user interface with the capability for construction and visualization. The former is often through a palette of drawing tools that can be used to draw the crude structure under investigation. Computational 'clean up' processes then turn this into a pseudo three-dimensional structure by generating spatial coordinates for each atom together with a connectivity list indicating which atoms should be bonded together. An alternative method is to obtain the molecular structure from another source, such as X-ray coordinates, and import it into the computational program. To date there is no universal file format for this although many programs will read coordinates from X-ray crystallographic information files, that have a .cif suffix, or from those described in the Brookhaven Protein Database format, with a .pdb suffix. Despite the wide use of these two file formats, different visualization software codes will often have a preferred variation of the file format making portability of structural information an ever-present problem for computational chemists working with more than one suite of programs.

After a structural model has been successfully created or imported the next step is to choose an appropriate force field to use. In practice this may not be an issue as the program accesses a default force field file; however, some programs contain several such files and may even have editable, or 'open force field', options that allow the user's own experimental data to be included. The user should be guided by the instruction manual when choosing a force field, or updating the parameters in one, as different force fields are more appropriate for some systems than others. Once a model has been created and a force field defined the overall energy needs to be minimized. In practice this is done by optimizing all the energy terms to give the lowest energy structure. The minimization routines themselves involve moving atoms to approach their ideal bond lengths and angles then analysing the effect this has on the total energy. All the functions that describe the atomic interactions are continuous and differentiable so the effect of moving every atom can be analysed from the derivatives of these functions. Moving any atom will change the structure's energy and once each move is complete a reassessment of the energy is possible. A decrease in energy indicates that the movement was in a beneficial direction and may be continued until the energy starts to rise again. This process is repeated for every atom until the apparent lowest energy geometry is obtained. The minimization programs are usually based on the Newton–Raphson method, which requires the calculation of both first and second derivative matrices. Different approximations (e.g. steepest descent or conjugate gradient) of these matrices can be used to speed up the computational process. The main problem in minimizing structures, as with all molecular modelling methods, is in identifying the differences between local and global energy minima. The geometry of the global energy minimum refers to the structure that has the absolute lowest steric energy. This particular geometry may never be observed in the laboratory, nevertheless, it is a valuable benchmark from which the energies of all other structures become relative. There are also many local energy minima with high-energy barriers that prevent them from adopting alternative geometries that are lower in energy, as indicated in Figure 4.7. The paradox inherent in this method is that if the structure

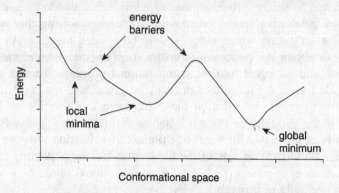

Figure 4.7   Local and global energy minima

is allowed to overcome high-energy barriers it is likely to miss some energy minima yet if it cannot overcome moderate energy barriers it may become stuck in a local minimum. This contradiction is usually resolved by a two-phase approach to minimization. Where the initial structure is far from a probable final geometry, as in a crude sketch created through the software package's graphical interface, its rough geometry is determined using a method that assumes high-energy barriers will be encountered. The second phase then proceeds with smaller iterative steps so that energy minima close to that of the improved structure are not overlooked.

Despite its apparent simplicity, and reliance on existing force fields to model potentially novel compounds or complexes, molecular mechanics is rapidly becoming the method of choice for theoreticians interested in accurate structure determination. Nowhere is the success of this technique more evident than in the prediction of protein structures. These molecules are far too large to be modelled by any other method, yet it is vital that the resulting geometry is accurate enough for its implications to be of value to molecular biologists. Although much of the analysis of a novel protein is down to homology mapping, where sections of the primary amino acid sequence are compared to known structures, molecular mechanics methods are still necessary to refine the overall geometry. As far as supramolecular chemistry is concerned, this simple technique is generally the most accurate and is the fastest method for generating a useful structural model of the system in question. The limitation of the method is that it cannot be used to derive thermodynamic parameters.

Ab initio methods

Ab initio calculations are so called because they assume nothing about the expected outcome of the experiment. No ideal bond lengths, angles or other parameters are supplied. The calculations rely solely on the solution of the Schrödinger equation for the molecular orbital, usually in the ground state, from which geometry and charge distribution are calculated. It is also possible to determine alternative molecular orbitals, such as excited states, and unusual spin states. The main problem with the *ab initio* technique is that the Schrödinger equation can only be solved accurately for H_2 and that solutions for molecules of any great complexity, particularly those involved in supramolecular chemistry, require that a large number of mathematical shortcuts are taken. Even then the calculations are computationally intensive. In the Hartree-Fock model, the most common method encountered, each atomic orbital is described in terms of a number of overlapping Gaussian functions that average to give a 'true' representation of the Schrödinger solution. The speed and accuracy of the overall calculation depends on how many Gaussian functions are invoked to describe s, p, d and f electrons. In addition to Hartree-Fock treatments other *ab initio* approaches, such as density functional methods, are also becoming popular due to more efficient calculation protocols that speed up simulations. At present the useful limit for *ab initio* calculations is around 200 atoms for commercial software running on a desktop computer. Coupled to this is the difficulty in guaranteeing that a molecular geometry derived using this approach is accurate. Fortunately it is usually possible to generate substantially accurate molecular geometries through a combination of conformational analysis and molecular mechanics. An *ab initio* interrogation of the resultant structure, a so-called single point energy calculation, can save vast amounts of computational resources as no high level iterative geometric calculations are required: the desired information (heat of formation, electron density dispersion, shape and disposition of the highest occupied and lowest unoccupied molecular orbitals) is calculated for just one geometry.

Semiempirical methods

As discussed above, calculations that employ the solutions to complex mathematical equations involving all the electrons in the system require no structural parameters to be supplied. These *ab initio* methods rely upon the correct interpretation of solutions to the Schrödinger wave equation to generate molecular orbitals that in turn predict properties such as equilibrium geometries, heats of formation and so on. The levels of complexity to which these calculations can now be undertaken are quite staggering yet there are two main drawbacks. First, the chemist has to have a good working knowledge of the mathematics behind the (often costly) software in order to use the most appropriate model and to correctly interpret the results. Second, and of greater relevance to supramolecular chemists, is the limit to the number of electrons that may be considered by this method. This leads to the limit of 200 atoms or so that can be considered at present and, on this scale, it is only useful for the simplest supramolecular systems. Without access to a supercomputer or a large parallel system this scale of simulation becomes the *de facto* limit for most chemists and equates to a calculation involving calixarene dimers or explicitly solvated crown ether complexes. While this may be a useful way to investigate the likely interactions of small guest molecules with small or medium sized hosts it rules out most multicomponent systems and solvated models. How then can we obtain quantum mechanical information, without which we cannot probe electron densities, transition states or an array of other properties, for supramolecular complexes? The answer is to consider semiempirical methods.

The main time-saving attribute of semiempirical methods is that they only consider valence electrons and assume localized atomic orbitals. This is the so-called Neglect of Diatomic Differential Overlap (NDDO) approximation. Under the NDDO approximation the number of electron–electron interactions scale as N^2 rather than N^4 where N is the number of mathematical functions used in the calculation. Other parameters derived from experiment, as used extensively in molecular mechanics methods, are incorporated to give further time-saving approximations. Despite these shortcuts it is still possible to generate well-founded models but, more importantly, the limit to semiempirical calculations on a desktop computer is increased to the order of 300 atoms. As with *ab initio* methods it is often wise to assume that a good molecular mechanics method will generate a suitably accurate structure for which a single-point energy calculation can be calculated. Other useful thermodynamic data can be determined using a combination of matrices calculated during the mechanics and energy phases of the simulation. These in turn can be used to generate atomic partial charges (useful in determining charge complementarity between host and guest), molecular orbitals, potential hydrogen bonds and the like. As with *ab initio* results these data may be visualized as described below.

Molecular dynamics methods

Molecular dynamics is the term used to describe a process where the effects of heating and cooling a molecule are simulated. At any particular temperature, above absolute zero, all molecules possess a certain amount of kinetic energy. The effects of this energy can be seen in bond vibrations and molecular motion. If enough energy is transferred to the molecule it may be able to overcome rotational or inversion barriers and adopt a different conformation to that of the starting structure. When simulations of heating are carried out for defined periods of time a large number of different conformers may be obtained for complex molecules. These conformers can be analysed in terms of relative abundance at particular temperatures or they may be allowed to 'cool' to ambient temperatures and have their structures optimized by molecular mechanics. The latter approach will usually generate fewer conformers, as some high-energy structures collapse into more favourable geometries at lower temperatures, but it may be more a realistic simulation in terms of the conformers available to the molecule under normal experimental conditions. The results can also be analysed as frequency counts for each conformer to give a statistical likelihood of each being present. Although this method cannot guarantee to find the global minimum geometry, nor will it necessarily find all the known conformers, a simulation run for long enough will generate a highly representative set of data for the conformers available to the target compound. Dynamics simulations are often carried out in tandem with Monte Carlo or conformational analyses. Each technique can be used to support the results of the other. For example, while the absolute energy values determined by the different simulations may not agree, the overall ranked order of the conformers detected would be expected to be broadly similar.

Periodic boundary conditions

Imagine that the test tube containing an aqueous solution of your host–guest complex actually contains a myriad of small boxes, or cells, each containing one host–guest complex and a specific number of water molecules. As with a crystallographic unit cell each repeat unit is the same and may be used to reconstitute the entire sample through replication in three dimensions. This is essentially what the periodic boundary model assumes. As each cell is identical to the next any molecule leaving the simulation through a wall will re-enter from the opposite wall as molecules cannot be lost or gained. The periodic boundary conditions keep the contents constant but also give consideration to the effects that neighbouring cells may have on the cell of interest. The simulation tends to be used on small molecules, such as [18]crown-6, in aqueous solution as the computational expense is quite high. The size of the cell is of key importance. Obviously the larger it is the more hosts, guests and solvent molecules can be included to give a more accurate representation of the system's complexity; however, there is a direct cost in the time it takes for these calculations. Most commercial software is limited in the range of solvents that can be included in the cell, with some restricted to water, and often only offer access to molecular mechanics or dynamic routines. As a result some of the most important supramolecular effects, such as electrostatics, dipoles and hydrogen bonding, are ignored by the calculations that merely generate sterically optimized geometries for the cell contents. Fortunately some programs now allow solvents other than water to be modelled explicitly and, most importantly, can split the analysis of the simulation using *ab initio* or semiempirical methods for the host–guest interactions and molecular mechanics for the remaining innocent chemical species and solvent molecules.

Such 'mixed mode calculations' have been shown to be of great value in studies of enzymes where the active site has been modelled using an *ab initio* approach and the remaining protein by simple mechanics. Given the rapid advances in computational speed and memory, together with faster molecular modelling algorithms, it is not unreasonable to expect that complex supramolecular systems will be modelled using periodic boundary conditions incorporating mixed *ab initio/* molecular mechanics methods in the near future.

Conformational analysis

Almost all molecules can exist in more than one conformation. With small or rigid conjugated systems it is possible to build all the major conformers and use their geometries as starting points to generate energetically minimized structures. The final energies can be compared and the one with the lowest steric energy should be the most stable. Indeed it may happen that all the starting geometries optimize to the same final structure. The problem faced by experimentalists attempting to model a compound or supramolecular complex of any substantial size is the sheer number of conformers that may exist. It is possible for the conformer that crystallizes from solution, and forms the basis for an X-ray structural determination, to be a minor constituent of the entire range of those that exist in solution.

In conformational analyses it is important to reduce the number of potential conformers while retaining those of most relevance to the experimentalist. So, although theoretically every bond can rotate through 360°, in practice it is assumed that multiple bonds are either *cis* or *trans* with respect to substituents generating torsion angles of 0° and 180° rather than the continuum of angles that intuitively should exist. Single bonds, through a similar logic, give rise to three possible conformers with torsion angles of $+120°$, $0°$ or $-120°$. Clearly conformers could also be generated at $60°$, $30°$ or even $1°$ increments; however, this increases the number of results and, more importantly, the computational time required.

The most obvious method to generate conformational data is to allow every bond in the molecule to rotate sequentially until all possible conformers have been obtained. Unfortunately this approach rapidly generates vast numbers of structures as the size of the molecule increases. Alternative methods use Monte Carlo methods or molecular dynamics to obtain a sample of structures. These methods are essentially random conformer generators and cannot guarantee that the lowest energy conformer has been detected. It is therefore less certain that the conformer closest to the global minimum geometry has been identified. Nevertheless these methods are often the only reasonable ways to generate the conformational range available to the molecule particularly when large structures are involved.

Once a group of conformers has been calculated for the molecule of interest two avenues are open for exploration. Some conformers will have extremely high energies as remote substituents are brought into close proximity and it will be clear that they are unlikely to reflect a high proportion of molecules in the gas or solution phase. It may be desirable to ignore these and pick the lowest energy conformer as the starting point for further geometry optimization calculations. Alternatively a more computationally intensive route would be to run a geometry optimization following the generation of each conformer. This approach removes high-energy conformers from the data set and, more importantly, determines if any of the conformers 'collapse' into the same optimum geometry or if several low energy conformers exist with diverse geometries. Systematic, Monte Carlo and molecular dynamics methods were all used in an analysis of [9]ane-S_3 and showed, quite surprisingly, that the lowest energy conformer had never been detected experimentally although many slightly higher energy conformers were known [1]. This illustrates the power of computational methods in their ability to predict experimentally unknown geometries.

Computational visualization

The sophistication of modern graphical user interfaces is now so great that it is possible to set up visualization methods that make the observer believe that he or she is manipulating giant three-dimensional molecules in mid-air. Using '3D glasses' and advanced projection systems it is possible to walk around, and even through, molecules as complex as enzymes and look in detail at important structural features. While this level of visualization, favoured by pharmaceutical companies and software vendors, is impressive it is perhaps more than the average experimentalist requires. Fortunately, even the simplest molecular visualization program running on an entry level computer is able to display molecules in various formats, from connectivities shown as sticks to space-filling representations, that illustrate host–guest complementarity. It is also easy to move and rotate molecules on screen so that particular features can be highlighted. This application is of particular use when submitting informative illustrations to journals or giving presentations. The power of molecular graphics is further enhanced by the superposition of physical properties such as hydrophobicities or electrostatics on three-dimensional depictions of molecules to visualize a 'fourth dimension'.

The insight gained from computer visualization can be illustrated with two examples from the sold state behaviour of calixarenes. In 2003 Stibor reported the structure of 25,26,27,28-tetrakis(1-propoxy)calix[4]arene-5,17-dicarboxylic acid [2]. The compound, being in the *cone* conformation, would be expected to form hydrogen-bonded dimers. Initial investigation of the crystal unit cell shows no such behaviour but if the structure is extended by the addition of another unit cell in the *c*-axis then dimers are immediately apparent. The symmetry inherent in the crystallographic space group makes such a relationship obvious to a well-trained crystallographer but not necessarily to the synthetic chemist. Fortunately, the availability of software to display crystallographic data allows the experimentalist to analyse solid-state structures for evidence of supramolecular assembly through hydrogen bonding and other interactions.

Another example comes from the Atwood group's work on water-soluble calix[4]arenes. In the late 1980s the crystal structures of several metal–4-sulphonatocalix[4]arene complexes were determined but it was the first of these, the tetrasodium salt, that fascinated the entire group. With the aid of an extremely expensive computer running visualization software that required tedious manual input of atomic coordinates it was possible to expand the number of unit cells and show that the solvated cations effectively formed channels that separated layers of calixarenes [3]. From this insight the analogy with the structure of clays was drawn, linking unnatural macrocycles back to the natural world and setting the scene for more complex supramolecular design based on 4-sulphonatocalixarenes.

A final example comes from a biological context. When, in 2001, McKinnon added to his breakthrough crystallographic report on the transmembrane potassium selectivity filter with a higher resolution analysis that gave an insight into the filter's mechanism [4], the paper was followed by Roux's computational analysis of the dynamic process [5]. The computational study was able to add information to the crystal structure, in particular, by 'alchemically' changing the potassium ions in the simulation to sodium ions and demonstrating that the pathway for the latter to traverse the ion channel was prohibitive.

[1] Conformational study of the macrocycle 1,4,7-trithiacyclononane in metal complexes, J. Beech, P. J. Cragg and M. G. B. Drew, *J. Chem. Soc., Dalton Trans.*, 1994, 719.

[2] Binding studies on the control of the conformation and self-assembly of a calix[4]-arenedicarboxylic acid through hydrogen bonding interactions, H. Miyaji, M. Dudic, J. H. R. Tucker, I. Prokes, M. E. Light, T. Gelbrich, M. B. Hursthouse, I. Stibor, P. Lhotak and L. Brammer, *J. Supramol. Chem.*, 2003, **15**, 385.

[3] Novel layer structure of sodium calix[4]arenesulfonate complexes – a class of organic clay mimics, A. W. Coleman, S. G. Bott, S. D. Morley, C. M. Means, K. D. Robinson, H. M. Zhang and J. L. Atwood, *Angew. Chem. Int. Ed. Engl.*, 1988, **27**, 1361.

[4] Chemistry of ion coordination and hydration revealed by a K^+ channel–Fab complex at 2.0 angstrom resolution, Y. F. Zhou, J. H. Morais-Cabral, A. Kaufman and R. MacKinnon, *Nature*, 2001, **414**, 43.

[5] Energetics of ion conduction through the K^+ channel, S. Bernèche and B. Roux, *Nature*, 2001, **414**, 73.

4.8   A Protocol for Supramolecular Computational Chemistry

Supramolecular systems, by definition, contain several molecules interacting through a combination of weak physical forces. To model the formation of these complexes each component and its environment must be considered as well as the supramolecular product. Complex formation is likely to involve desolvation (or partial desolvation) of each component and, if a flexible or macrocyclic compound is involved, consideration of its preferred geometry. The latter may be achieved through analysis of structural data or by conformational analysis. The former is best approached using a Born–Haber style analysis that can be used for a general host–guest system in which the host is a flexible macrocycle; this is illustrated in Figure 4.8. Simulations, probably at molecular mechanics level, can be carried out on the solvated host and guest components together with the solvated host–guest complex. This is partly to give a more accurate visual experience (the experimentalist can 'see' the compounds in solution) and partly to determine if the solvent molecules affect the geometries of the individual components or complex (which they often do). Solvent molecules are then removed from the simulations and high-level single point energy calculations undertaken on the remaining components and complex, respectively, to find ΔG or ΔH (difference between free energies or enthalpies of formation). In addition to these, two further simulations are required to assess the effect of host–guest complexation. Using whichever method is most appropriate (Monte Carlo, conformational searching or molecular dynamics), the lowest energy geometries for the host and guest are determined and, using high-level calculations, thermodynamic data generated. This is repeated for the host–guest complex. The outcome of these calculations is that the effect of solvent is explicitly modelled for every species in the $H + G \leftrightarrows H \cdot G$ equilibrium and the energies (ΔG or ΔH) are known for host and guest (in their most stable conformers) and host–guest complex (in its most stable conformer). The effects of gas and solution phase complexation then can be examined.

For overall stability the free energy of the complex must be more negative that that of the sum of the components. Most computational packages are able to calculate ΔH rather than ΔG but, as the number of molecules in the complex is likely to be less than that of the components, the ΔS (difference in entropy) term

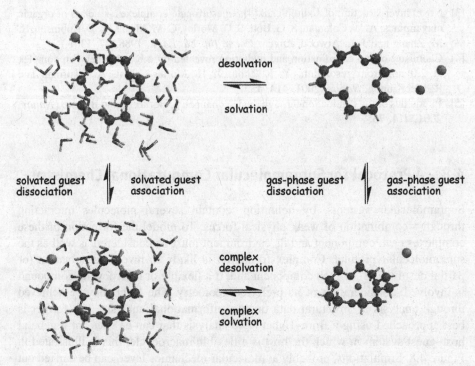

Figure 4.8 A protocol for supramolecular chemistry *in silico*

will usually be smaller for the complex than the components. ΔH then becomes a good indicator of supramolecular stability. This measure of supramolecular stability may be summarized as:

$$\Delta H \text{ (complex)} < \Sigma \; \Delta H \text{ (components)}$$

For more complicated systems, such as multicomponent supramolecules and biomolecules, it may be more appropriate to quote the relative difference in steric energy as a measure of stability. This can be quickly calculated by molecular mechanics methods.

4.9   Examples of *In Silico* Supramolecular Chemistry

Use of molecular mechanics: [15]Crown-5 and [18]crown-6 as complexants for alkali metals

To return to the question posed in Section 4.6, which is more likely to form a complex with [15]crown-5 in aqueous solution, Na^+ or K^+? What about

[18]crown-6? Based on well-known examples, Na^+ should be a better size match for the [15]crown-5 cavity and K^+ for [18]crown-6. However, experimental results give log$K$ values of 0.7 and 0.8 for Na^+ complexes with [15]crown-5 and [18]crown-6, respectively. More surprisingly the log$K$ values of the K^+ complexes with [15]crown-5 and [18]crown-6 are 0.75 and 2.1, respectively. This indicates that a greater affinity is exhibited for K^+ complexes over Na^+ complexes with either crown. There is also little apparent attraction between Na^+ and the crowns based upon cavity size. Can these results be replicated by simple computational experiments?

To illustrate this models of the solvated alkali metal chlorides, ligands and complexes were generated using a drawing program within the computational chemistry software suite. It is necessary to compute the energies of the ligand in its most stable conformer, the energies of the metal salts, and the energies of the complexes. Ordinarily this would be a simple matter of generating the lowest energy conformers and noting their relative energies but in this case the process is complicated because the solvent is likely to influence the most stable conformation of the ligand. One approach is to invoke a calculation that implicitly includes the effects of solvent but such an approach would ignore the specific effects that may result from hydrogen bonding. Here the structures of the lowest energy conformers were calculated by a molecular mechanics-based energy optimization of the complexes surrounded by 55 water molecules. An example of one such system is shown in Figure 4.9. Solvent molecules were explicitly included to represent

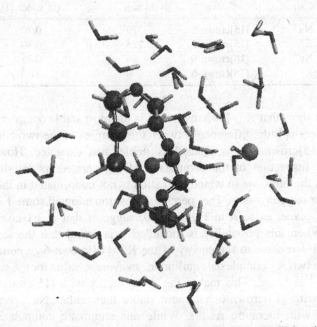

Figure 4.9 A crown ether–metal complex simulation using explicit solvent

molar solutions and to allow visual inspection of solvent-related effects in the low energy conformers. A separate calculation was made for the alkali metal chlorides. Final analysis based upon the solvated complexes would be a formidable task on a desktop or laptop computer so, to speed up the process, solvent molecules were removed from all simulations prior to calculating the overall energies of the ligands, ion pairs and complexes. In any event, the energy due to solvents would have merely been additive to that of each simulation and would have to be removed to make a true comparison between models. Single-point energies, that is to say energies of the resulting structures without further geometry optimization, were calculated, again using the low molecular mechanics level of theory. This gave relative steric energies for molecules, in geometries that had been influenced by solvent effects but avoided having to calculate the energy contributions from the 55 water molecules. The two sets of supramolecular binding events described below can then be tabulated and compared to literature data (Table 4.2).

$$[15]\text{crown-}5_{(aq)} + MCl_{(aq)} \rightarrow M^+ \cdot [15]\text{crown-}5_{(aq)} + Cl^-_{(aq)}$$

$$[18]\text{crown-}6_{(aq)} + + MCl_{(aq)} \rightarrow M^+ \cdot [18]\text{crown-}6_{(aq)} + Cl^-_{(aq)}$$

$$M = Na \text{ and } K$$

Table 4.2 Cation–crown ether complex formation

Cation	Crown	Relative energy (kJ mol^{-1})	Calculated log K (in water [1])
Na$^+$	[15]crown-5	70	0.70
K$^+$	[15]crown-5	125	0.80
Na$^+$	[18]crown-6	80	0.75
K$^+$	[18]crown-6	0	2.10

The results show that $K^+ \cdot [18]\text{crown-}6_{(aq)}$ is the most stable complex, as expected, and that there is little difference between the energies of the two Na$^+$ complexes. The $K^+ \cdot [15]\text{crown-}5_{(aq)}$ is much less stable than expected. How can this be explained? Inspection of the computationally generated model shows that this complex is the only one in which the cation is not coordinated in the plane of the crown ether oxygen atoms. The 'perching' position adopted some 1.4 Å (0.14 nm) above this plane, as seen in Figure 4.10, suggests that a 2:1 complex may be feasible. When this possibility is modelled it is found that the steric energy is -55 kJ mol^{-1} relative to the energy of the $K^+ \cdot [18]\text{crown-}6_{(aq)}$ complex. Assuming that the two K$^+$ complexes equilibrate, an average value for the steric energy of 35 kJ mol^{-1} is found. This makes the K$^+$ complex with [15]crown-5 less stable than that with [18]crown-6 but more stable than either Na$^+$ complex, and in agreement with literature results. While this simplistic computational protocol does not match the literature perfectly, specifically the reversal of Na$^+$ stabilities

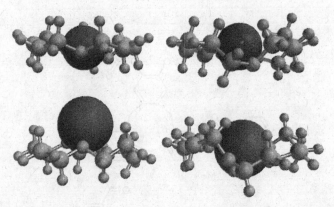

Figure 4.10 Simulated binding modes for crown ether–metal complexes: Na^+-[15]crown-5 (top left), Na^+-[18]crown-6 (top right), K^+-[15]crown-5 (bottom left), K^+-[18]crown-6 (bottom right)

with the crowns, it nevertheless enables a sensible hypothesis to be reached concerning the likely form of the species in solution. Visual inspection indicates that 1:1 complexes are likely to form when the cation sits in the plane of the crown. The only combination that could form a 2:1 complex is K^+ with [15]crown-5 as the ligand and, in doing so, it yields an explanation for the anomalous binding constant for that species. Furthermore, a survey of crystal structures incorporating K^+ and [15]crown-5 quickly reveals that this motif is common in the solid state.

Use of conformational analysis: self-complementarity of cyclotriveratrylene–molecular capsules

In 2004 Hardie published the synthesis and X-ray crystal structure of a cyclotriveratrylene derivative, tris(isonicotinyl)cyclotriguaiacylene (Figure 4.11), that formed solvent-free dimers when crystallized from dichloromethane following vapour diffusion of hexane [2]. The dimer not only exhibited interdigitation of pyridine substituents but also implied the existence of weak π–π interactions between the pyridine rings. Is it possible to replicate this structure computationally? As a starting point it is easy to create the macrocycle with a suitable building interface available with most molecular modelling software packages. To determine the likelihood of self-complementary complexes to form capsules the energies of both the most stable conformer and the one that is actually found in the dimer must be determined. The latter should form the most stable self-assembled complex as each aromatic ring can engage in π–π stacking. Fortunately, the crystal structure shows no solvent inclusion that allows gas phase methods to be used and offers a substantial reduction in the computational time required.

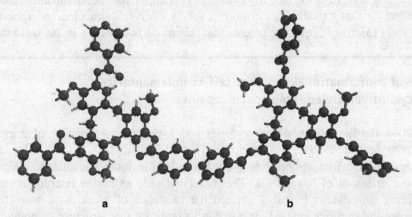

Figure 4.11   Tris(isonicotinyl)cyclotriguaiacylene

Initially a full conformational search is made for the molecule. This generates a number of conformers with slightly different steric energies. The lowest energy conformer, illustrated in Figure 4.12, has pyridine rings in unsuitable positions for interdigitation. Inspection of the conformers generated reveals an example with

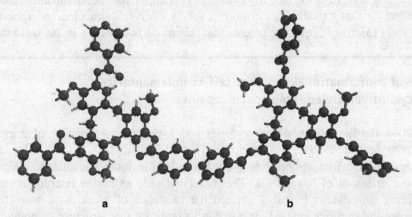

a   b

Figure 4.12   Tris(isonicotinyl)cyclotriguaiacylene conformers with the lowest energy (left) and the optimum geometry for capsule formation (right)

substituents perpendicular to the aromatic rings in the parent macrocycle that might dimerize (Figure 4.12). This conformer is chosen and a copy 'docked' into an appropriate position to induce capsule formation. Following molecular mechanics-based structural optimization a stable dimer forms (Figure 4.13). Further single point energy calculations using semiempirical methods give ΔH

Figure 4.13 Calculated structure of a tris(isonicotinyl)cyclotriguaiacylene dimer (side and top views)

values of $-293\,kJ\,mol^{-1}$ for the dimer and $-133\,kJ\,mol^{-1}$ for the lowest energy conformer of the monomer. Using the requirement for stability, ΔH (dimer) $<$ $\Sigma\,\Delta H$ (monomer), the dimer is predicted to be more stable by $27\,kJ\,mol^{-1}$. In addition the distances between pyridine centroids involved in apparent π–π interactions are calculated to be 3.70 Å (0.37 nm), somewhat shorter than the value of 4.20 Å (0.42 nm) found in the crystal structure but, nevertheless, in good agreement with expected value of 3.80 Å (0.38 nm) [3].

Use of *ab initio* methods: relative stabilities of calix[4]arenes

Calix[4]arenes can adopt four conformers: *cone*, *partial cone*, *1,2-alternate* and *1,3-alternate*. In the case of 4-*t*-butylcalix[4]arene the interconversion between conformers is possible above 60 °C or so. During heating reagents can be added to influence the final conformer of the calixarene once the solution cools. The addition of caesium yields the *1,3-alternate* conformer predominantly; no added cation gives the stable, hydrogen-bonded *cone* conformer. The influence of alkali metals on the conformers of the calix[4]arene anions remains a current interest [4]. Can molecular modelling predict the relative stabilities of these four conformers? Two approaches are informative and both model the desolvated, or gas, phase. The first requires a conformational search, followed by geometry optimization, to generate the different conformers and calculate their relative steric energies. The second uses conformers generated by the experimentalist. Both methods are best suited to molecular mechanics treatments which are very accurate at predicting ideal bond lengths and angles of these simple compounds. In this example it is easiest to generate a cone conformer then break, rotate and reform bonds between bridging methylene groups to generate the other three conformers. The main reason for this is that many conformational analysis programs will not interconvert calix[4]arene conformers using default parameters. It may be necessary to raise the temperature of the simulation or invoke a Monte Carlo approach to do so.

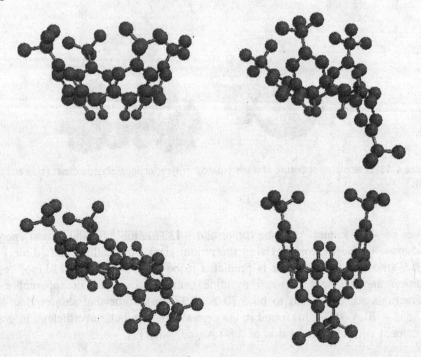

Figure 4.14 Calix[4]arene conformers (hydrogen atoms removed for clarity); *cone* (top left), *partial cone* (top right), *1,2-alternate* (bottom left), *1,3-alternate* (bottom right)

Nonetheless, with only four conformers, it is easier to generate them manually than to apply these searching methods (Figure 4.14). Once the steric energies of the different conformers have been calculated, using geometry optimized structures, it is possible to rank them.

The relative energies shown in Table 4.3 are in agreement with experimental data that show the *cone* form as the most stable conformer. Why is this? Overlaying an electron density map, calculated from the initial geometry using *ab initio* methods, on a *cone* structure shows that the lower rim phenolic groups

Table 4.3 Relative energies computed for *4-tert*-butylcalix[4]arene conformers

Conformer	Relative energy (kJ mol^{-1})
Cone	0
Partial cone	54
1,2-Alternate	69
1,3-Alternate	92

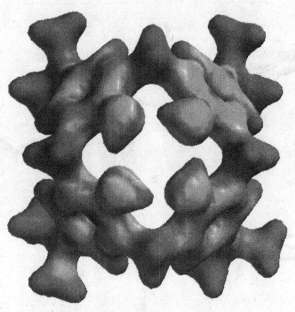

Figure 4.15 An electrostatic bond density map showing strong hydrogen-bonding in *cone-4-t-butylcalix[4]arene*

form a strong hydrogen-bonded network that must be disrupted before any conformational interconversion can begin (Figure 4.15). Simulations of de-*tert*-butylated calix[4]arene indicate the same rank order of the conformers, with the *cone* as the most stable, followed by the *partial cone* ($+42\,kJ\,mol^{-1}$), the *1,2-alternate* ($+48\,kJ\,mol^{-1}$) and *1,3-alternate* ($+71\,kJ\,mol^{-1}$). It is proposed that the *partial cone* acts as a route to either of the alternate conformations and it is itself relatively unstable [5]. If the *cone* is so stable, how come caesium can force it into the least favourable conformer? Modelling the caesium salts of both conformers and calculating their associated steric energies should be able to give a rapid answer.

Use of conformational analysis to determine energy barriers: through-the-annulus rotation of oxacalix[3]arenes

As with other members of the calixarene family, oxacalix[3]arenes have the ability to rotate their phenolic rings through the annulus of the macrocycle as shown by proton NMR in which the hydrogen atoms in the ethoxy bridges appear as sharp singlets. If the phenolic moieties are treated with alkyl halides in the presence of base then the respective trisubstituted alkyl ethers are formed. Increasing the steric bulk of the ether derivatives makes through-the-annulus rotation progressively

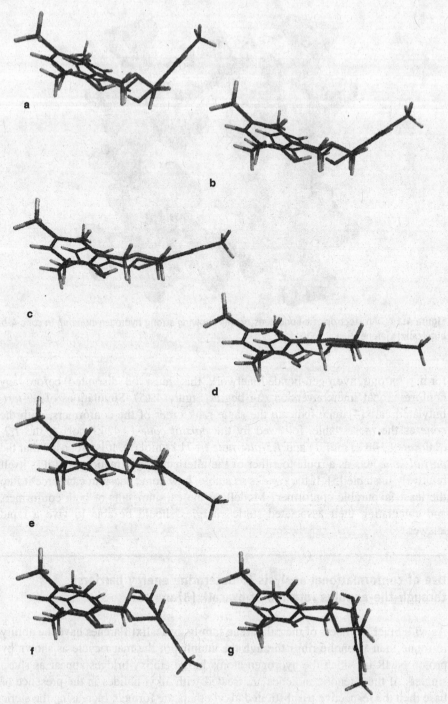

Figure 4.16 Through-the-annulus rotation of an oxacalix[3]arene from *cone* (a) to *partial cone* (g)

more difficult leaving the macrocycles 'frozen out' in either the *cone* or *partial cone* conformer as evidenced by splitting of the alkyl bridge signal in the proton NMR. The temperature at which free rotation is inhibited, the coalescence temperature, observed for these systems is related directly to the energy required to force the lower rim substituents through the annulus. It is possible to replicate this phenomenon by rotating a phenolic ring through the annulus of an oxacalix[3]arene and optimizing the geometry at each stage? The example given examines the effect of a 180° rotation by one phenolic ring in 10° increments through the annulus of 4-methyloxacalix[3]arene where the lower rim substituents are increased in steric bulk from hydrogen to methyl, ethyl, *n*-propyl and finally *n*-butyl groups. Snapshots of through-the-annulus rotation for the parent compound are shown in Figure 4.16. The resulting graph of torsion angle against energy illustrates the effect. Results show that, at a simulated temperature of 25 °C, the parent compound has a barrier to rotation of *ca*. 20 kJ mol^{-1} (illustrated in Figure 4.17), the tri(methyl ether) derivative a barrier of *ca*. 100 kJ mol^{-1}, and the

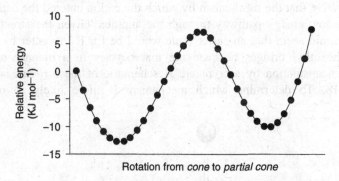

Figure 4.17   An energy profile of oxacalix[4]arene conformer interconversion

more bulky ethers achieve highly strained geometries (*ca*. 65 kJ mol^{-1} above the energy of the geometry optimized conformer) but do not rotate through the annulus. The difficulty of through-the-annulus rotation with increasing bulk correlates well with literature data where coalescence values lower than −50 °C for parent and methyl derivatives, 50 °C for ethyl and 100 °C for propyl derivatives have been reported in variable temperature NMR experiments [6]. More bulky groups are frozen in either the cone or partial cone conformer. Where the power of molecular simulations becomes apparent is in the ability to chart torsion angles with energy and changes in geometry. For example, snapshots of the parent compound show that the first barrier to rotation is the phenolic hydrogen-bonding network. When a phenolic ring rotates, its $O \cdots O-H$ interaction with an adjacent phenolic proton increases steadily with the rotation angle whilst the latter starts to interact more strongly with an ethereal oxygen. Thus at the first energy minimum the distances are 2.13 Å (0.213 nm) and 1.91 Å (0.191 nm) but at the second they are 3.13 Å (0.313 nm) and 1.61 Å (0.161 nm) for the O—H···O—H and O—H···O$_{(ether)}$ interactions, respectively. Interestingly, the coalescence temperatures reported for the analogous cyclophanes (where the interphenolic propyl bridges contain no heteroatoms) are higher, presumably because, in addition to the greater steric bulk of the bridging CH$_2$ over O, there is no pathway by which energy from a breaking hydrogen bond may be progressively redirected into the strengthening of another.

Use of molecular mechanics to model a dynamic process: Bis(calix[4]arene) complexes of K$^+$

In 1997 Beer reported a remarkable bis(calix[4]arene) compound together with X-ray crystal structures of the ligand and its K$^+$ complex [7]. For the system to be considered as a model for the selectivity filter in a transmembrane K$^+$ channel it was imperative that the mechanism by which the cation entered the central cavity involved a low energy pathway through the annulus. Given the structure of the ligand it would seem that an easier route would be for K$^+$ to enter by squeezing through the ethyl bridges between the macrocycles in a manner reminiscent of metal encapsulation by a cryptand. A schematic of these routes is shown in Figure 4.18. To determine which mechanism is more likely to occur, two

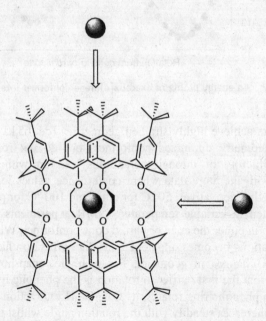

Figure 4.18　Routes to cation encapsulation by a biscalixtube

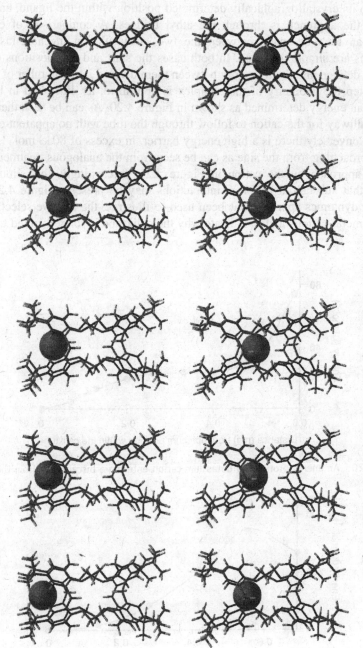

Figure 4.19 Sequential snapshots of a potassium cation entering a biscalixtube from the end

simulations are necessary: one in which the cation is directed to move through the annulus to its crystallographically determined position within the ligand, and one in which the approach is through the ethyl bridges. A comparison of energy barriers can then be made between the two encapsulation mechanisms. The simulations are simple to set up. In both cases the start and end positions of the cation are defined and the trajectory between them divided into a number of steps. At each step the geometry of the complex is optimized, as illustrated in Figure 4.19, and an energy determined as shown in Figure 4.20. As can be seen there is a smooth pathway for the cation to follow through the tube with no apparent energy barriers. Conversely there is a high energy barrier, in excess of $80\,\text{kJ mol}^{-1}$, for a cation approaching from the side as can be seen from the analogous treatment of a sideways approach by the cation in Figure 4.21. Interestingly, this simulation indicates that substantial cation–π interactions may be present (Figure 4.22). A molecular dynamics approach has been used to illustrate the relative selectivities for alkali metal cations and to indicate why the calixtube is so well suited to bind K^+ [8].

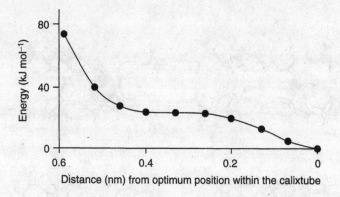

Figure 4.20 An energy profile of a potassium cation entering a biscalixtube from the end

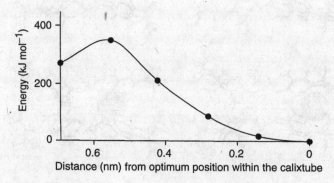

Figure 4.21 An energy profile of a potassium cation entering a biscalixtube from the side

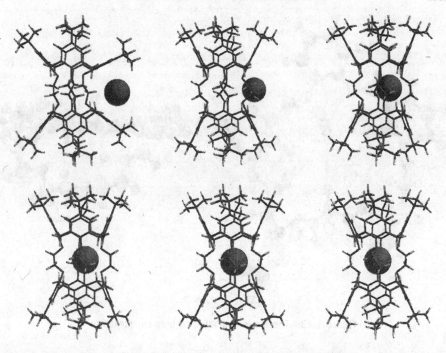

Figure 4.22 Sequential snapshots of a potassium cation entering a biscalixtube from the side

Visualization of designer macrocycles

One of the main endeavours of supramolecular chemistry must surely be the synthesis of compounds for specific purposes based not upon chance but upon a clear understanding of the structural elements that are required to make 'supramolecules'. Thus a supramolecular chemist may analyse a problem related to the detection of a specific analyte in terms of the complementary functionality necessary to bind the target. This will undoubtedly involve encapsulation, implying that a specific size and rigidity of macrocycle will form the basis of the detector, and the type of binding sites required within the macrocycle. Typically these will include hydrogen bond donors and acceptors and aromatic groups with the potential to be involved in π–π stacking. Three examples highlight the concept of supramolecular chemistry by design: Reinhoudt's urea receptor [9,10], Hamilton's barbiturate receptors [11,12] and Bell's creatinine receptor [13]. Each case illustrates the complex supramolecular principles that must be considered when designing a molecule that must recognize a particular guest in the presence of many potential interferents. All three are excellent examples of supramolecular chemistry by design.

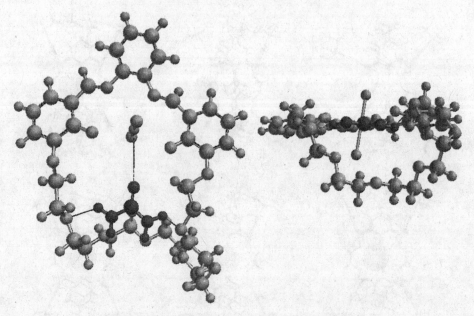

Figure 4.23   Interactions leading to urea complexation

Urea is a relatively simple molecule containing amine and ketone functionality. It is produced by the body as a method of removing excess nitrogen and is excreted in urine. It should be possible therefore to design a ligand that complements urea's functionality. A simple analysis shows that the three functional groups form a divergent trigonal planar shape and that the amines could interact well with oxygen atoms in [18]crown-6. Unfortunately the ketone would repel crown ether oxygens. While a protonated azacrown ether may form a stable complex with urea, the ligand is susceptible to interference from a variety of sources including water and alkali metals. The Reinhoudt group's innovation was to use a salphen-bound uranyl group as the ketone binding site [9]. Generally metal–salphen complexes contain a square planar or octahedral metal chelated to two nitrogens and two oxygens from the ligand. The uranyl–salphen complex is unusual in that uranium is so large that the metal adopts a pentagonal bipyramidal geometry, making a fifth equatorial binding site available for ligation. The Reinhoudt group incorporated the uranyl–salphen complex into [18]crown-6 to introduce a metallic binding site for the urea carbonyl group [10]. The effectiveness of this concept was proven by the crystal structure of the ligand-urea complex in which the guest was bound exactly as anticipated. A simulation of the complex, shown in Figure 4.23, replicates both the guest binding and the 'wrapped around' geometry of the crown section of the ligand.

In the 1980s the Hamilton group applied the idea of complementary binding sites to the detect barbiturate derivatives. Barbiturates have alternating ketone and amine functionality that diverges from a cyclohexyl ring. To bind them the group first designed a receptor containing a rigid array of six complementary hydrogen bond donors and acceptors together with aromatic groups that form CH–π interactions with the terminal methyl groups of barbital [11]. Improvements on this original strategy included replacing the diphenylmethane bridge with a naphthalene spacer that stacked above the barbiturate core. As with the urea complex mentioned above, the X-ray structure showed that the aromatic spacer stacked effectively over the guest. Again, this is replicated in a simple molecular mechanics simulation of the complex, shown in Figure 4.24. The binding constants

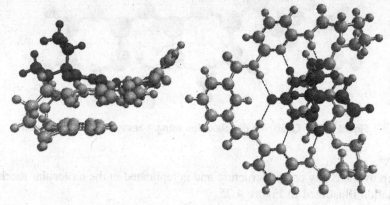

Figure 4.24 Barbiturate complexation by both complementary hydrogen-donor and acceptor groups and π–π interactions

for the complexes ranged from 10^2 M^{-1} to a highly respectable 10^6 M^{-1}, indicating that specificity could be achieved for the target analyte through careful design of the receptor [12].

In the 1990s the Bell group used the 'hexagonal lattice' approach to design a creatinine-specific receptor that could be used as a hospital-based assay for that analyte. Creatinine is an indicator of kidney function and, in cases where renal failure is suspected, a rapid method of detection would be of immense value as appropriate medical intervention could be requested in a timely fashion. The difficulty with sensing creatinine hinges on its structure: it is a five-membered ring containing amine, ketone and *N*-methyl functionality. Through a combination of hydrogen-bond donors and acceptors part of the creatinine functionality could be recognized and, using a bifurcated hydrogen bond, the molecule bound by the hexagonal lattice. In addition to accurately recognizing the metabolite, the receptor's fluorescence was quenched upon creatinine. Unfortunately, the receptor was insoluble in water and fluorescence was only quenched in methanolic solution. After a slight, though by no means trivial, modification to the receptor synthesis a version was prepared that bound creatinine in water-saturated dichloromethane that underwent a chromogenic response [13]. The host–guest interaction was

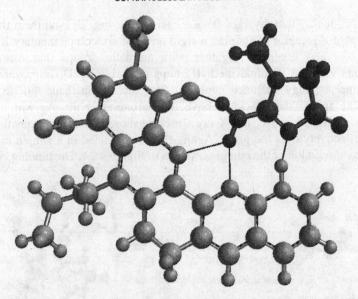

Figure 4.25 Creatinine complexation using a hexagonal lattice receptor

proven by the X-ray crystal structure and is replicated in the molecular mechanics simulation illustrated in Figure 4.25.

[1] Thermodynamic and kinetic data for macrocycle interaction with cations and anions, R. M. Izatt, K. Pawlak, J. S. Bradshaw and R. L. Bruening, *Chem. Rev.*, 1991, **91**, 1721.

[2] Interwoven 2-D coordination network prepared from the molecular host tris(iso-nicotinoyl)cyclotriguaiacylene and silver(I) cobalt(III) bis(dicarbollide), M. J. Hardie and C. J. Sumby, *Inorg. Chem.*, 2004, **43**, 6872.

[3] A critical account on $\pi-\pi$ stacking in metal complexes with aromatic nitrogen-containing ligands, C. Janiak, *J. Chem. Soc., Dalton Trans.*, 2000, 3885.

[4] Synthesis, structures, and conformational characteristics of calixarene monoanions and dianions, T. A. Hanna, L. Liu, A. M. Angeles-Boza, X. Kou, C. D. Gutsche, K. Ejsmont, W. H. Watson, L. N. Zakharov, C. D. Incarvito and A. L. Rheingold, *J. Am. Chem. Soc.*, 2003, **125**, 6228.

[5] Molecular modeling of calixarenes and their host–guest complexes, F. C. J. M. van Veggel, in *Calixarenes in Action*, L. Mandolini and R. Ungaro, Eds., Imperial College Press, London, 2000.

[6] Conformational isomerism in and binding-properties to alkali-metals and an ammonium salt of *O*-alkylated homooxacalix[3]arenes, K. Araki, K. Inada, H. Otsuka and S. Shinkai, *Tetrahedron*, 1993, **49**, 9465.

[7] Calix[4]tube: a tubular receptor with remarkable potassium ion selectivity, P. Schmitt, P. D. Beer, M. G. B. Drew and P. D. Sheen, *Angew. Chem. Int. Ed. Engl.*, 1997, **36**, 1840.

[8] Selectivity of calix[4]tubes towards metal ions: a molecular dynamics study, V. Felix, S. E. Matthews, P. D. Beer and M. G. B. Drew, *Phys. Chem. Chem. Phys.*, 2002, **4**, 3849.

[9] Cocomplexation of urea and UO_2^{2+} in a Schiff-base macrocycle – a mimic of an enzyme binding-site, C. J. van Steveren, D. Fenton, D. N. Reinhoudt, J. van Erden and S. Harkema, *J. Am. Chem. Soc.*, 1987, **109**, 3456.

[10] Cocomplexation of neutral guests and electrophilic metal cations in synthetic macrocyclic hosts, C. J. van Staveren, J. van Erden, F. C. J. M van Veggel, S. Harkema and D. N. Reinhoudt, *J. Am. Chem. Soc.*, 1988, **110**, 4994.

[11] Molecular recognition of biologically interesting substrates: synthesis of an artificial receptor for barbiturates employing six hydrogen bonds, S.-K. Chang and A. D. Hamilton, *J. Am. Chem. Soc.*, 1988, **110**, 1318.

[12] Hydrogen-bonding and molecular recognition – synthetic, complexation, and structural studies on barbiturate binding to an artificial receptor, S. K. Chang, D. Vanengen, E. Fan and A. D. Hamilton, *J. Am. Chem. Soc.*, 1991, **113**, 7640.

[13] Detection of creatinine by a designed receptor, T. W. Bell, Z. Hou, Y. Luo, M. G. B. Drew, E. Chapoteau, B. P. Czech and A. Kumar, *Science*, 1995, **269**, 671.

5

Supramolecular Phenomena

5.1 Clathrates

Interstitial molecular complexes, or clathrates, were first observed by Humphrey Davy who described the solid gas clathrate formed by chlorine inclusion in ice [1]. The nature of this inclusion phenomenon was further studied by Michael Faraday who demonstrated that chlorine could be liberated from the 'chlorine hydrate' [2]. This was later shown by Linus Pauling to be $(H_2O)_{7^{2}/_{3}} \cdot Cl_2$ [3]. The clathrate phenomenon has also been observed for many other small molecules including the noble gases. One of the more important examples is the formation of methane hydrates in gas pipelines. When methane enters a pipe under pressure at low temperatures it can cause crystallization of any water that is present. The resulting clathrate is comprised of methane molecules held within a hydrogen-bonded network of frozen water molecules. Methane, unusually for a small organic molecule with no functional groups, actually acts as a template around which low density ice forms. In the absence of guest molecules the more common compact ice structure results. As might be expected both the formation and inhibition of gas clathrate hydrates is of great industrial importance. Methane clathrates, found mostly in ocean floors and areas of permafrost, are attractive sources of methane itself due to their global abundance and availability, with perhaps as much as 10^{16} kg present in the Earth's crust. Their presence was first noticed in pipelines in the 1930s and their nature determined in the 1960s. Although their presence may cause alarm, the clathrates are not necessarily hazardous in themselves. Unless they undergo dramatic changes in pressure or temperature, or are ignited, their decomposition around 0 °C is not explosive. Problems do arise when gas hydrates form in natural gas pipelines and cause blockages that can lead to catastrophic failure. To this end the structures of the hydrates, such as that shown in Figure 5.1, remain an area of supramolecular investigation [4]. As will be seen the clathrate phenomenon is not restricted to the solid phase.

A Practical Guide to Supramolecular Chemistry   Peter J. Cragg
© 2005 John Wiley & Sons, Ltd

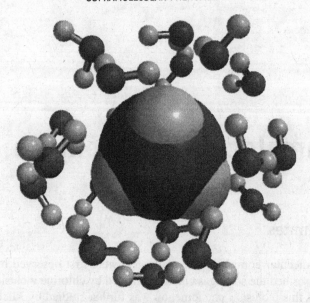

Figure 5.1 Simulation of a methane hydrate

[1] On a combination of oxymuriatic gas and oxygene gas, H. Davy, *Philos. Trans. Royal Soc. London*, 1811, **101**, 155.

[2] On hydrate of chlorine, M. Faraday, *Quart. J. Sci.*, 1823, **15**, 71.

[3] The structure of chlorine hydrate, L. Pauling and R. E. Marsh, *Proc. Natl. Acad. Sci. U S A*, 1952, **38**, 112.

[4] Gas hydrate single-crystal structure analyses, M. T. Kirchner, R. Boese, W. E. Billups and L. R. Norman, *J. Am. Chem. Soc.*, 2004, **126**, 9407.

5.2 Stabilization of Cation--Anion Pairs by Crown Ethers: Liquid Clathrates

Liquid clathrates, analogues of the gas clathrates but present in the liquid phase, are unusual two phase systems in which an upper layer of solvent lies above a denser layer of solvent saturated with ionic species. Some interesting examples are found in Atwood's work. Initially a salt, such as sodium chloride, is added to an aromatic solvent in which it is insoluble. Another reagent, an aluminium alkyl in one example, is added and the salt is solublized to form a dense phase [1] which may be interpreted as:

$$MX + 2\,AlR_3 + \text{excess aromatic} \rightarrow M^+[Al_2R_6 \cdot X]^- \cdot n\,\text{aromatic}$$

It was later discovered that addition of a stoichiometric quantity of an appropriate crown ether aided the formation of the complex salts [2,3] through encapsulation of the metal cation:

$$\text{crown} + 2\,\text{AlR}_3 + \text{excess aromatic} \rightarrow \text{crown} \cdot [\text{AlR}_3]_2 \cdot n \text{ aromatic}$$

$$\text{crown} \cdot [\text{AlR}_3]_2 \cdot n \text{ aromatic} + \text{MX} \rightarrow [\text{crown} \cdot \text{M}]^+ [\text{Al}_2\text{R}_6 \cdot \text{X}]^- \cdot n \text{ aromatic}$$

In an extension to this work it was found that simple metal halides (and some carbonyls) could also form liquid clathrates with [18]crown-6 in toluene or benzene when HCl was bubbled through the mixture, but only in the presence of trace water (or 'fortuitous water', as it became known within the group) [4]. These systems also generated hydronium ions that were stabilized by the crown ether and crystallized from solution.

$$\text{crown} + 2\,\text{HCl} + \text{H}_2\text{O} + \text{aromatic} \rightarrow [\text{crown} \cdot \text{H}_3\text{O}]^+ [\text{HCl}_2]^- \cdot n \text{ aromatic}$$

$$[\text{crown} \cdot \text{H}_3\text{O}]^+ [\text{HCl}_2]^- \cdot n \text{ aromatic} + \text{MCl} \rightarrow [\text{crown} \cdot \text{M}]^+ [\text{HCl}_2]^- \cdot n \text{ aromatic}$$

Formation of the clathrate is easy to follow visually as two phases separate in the reaction vessel, as shown in the schematic in Figure 5.2. Confirmation may be

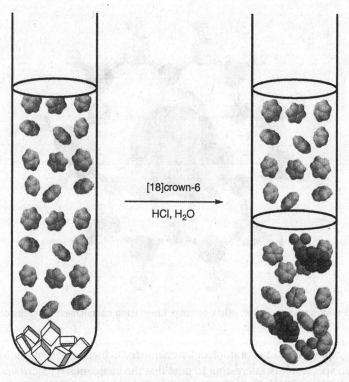

[18]crown-6

HCl, H₂O

Figure 5.2   Formation of a liquid clathrate from a metal salt and [18]crown-6 in benzene

obtained by ^{1}H nuclear magnetic resonance (NMR): a sample from the upper phase will exhibit signals at 7.2 and 2.3 ppm in the ratio of 5:3, diagnostic of toluene (or a signal at 7.1 ppm for six benzene protons). A sample from the lower phase will also contain signals for toluene, however, a peak at 3.7 ppm equivalent to 24 protons will also appear for [18]crown-6. The ratio of solvent to crown can be determined by comparing the relative intensities of the aromatic and crown peaks. Generally the crown species is heavily solvated with ratios of 15:1 (solvent:crown) being typical. If the lower phase is allowed to crystallize the resulting solid clathrates contain far less solvent, usually only 0.5 to 2 equivalents, as determined by ^{1}H NMR and X-ray crystallography [4].

An important result from this procedure is the ability to isolate the ([18]crown-$6 \cdot H_3O)^+$ cation in a non-aqueous solvent [5]. Although early spectroscopic studies pointed to the presence of H_3O^+ bound within a macrocyclic cavity [6] the X-ray structure was invaluable in illustrating the physical nature of the strongly hydrogen-bonded interactions between host and guest. Consideration of valence shell electron pair repulsion theory predicts a tripodal geometry for the oxonium cation yet crystal structures show that the oxonium oxygen lies less than 0.1 Å (0.01 nm) from the plane made by the closest three oxygen atoms in [18]crown-6. This is also observed in simulations as illustrated in Figure 5.3. The role that X-ray

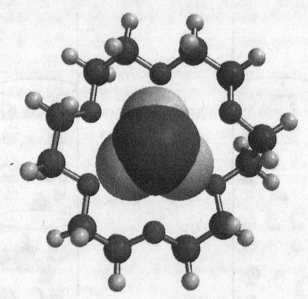

Figure 5.3 An oxonium--crown ether complex illustrating complementary threefold symmetry

crystallography plays in supramolecular chemistry is highlighted by the structure of this crown species. It is interesting to note that the analogous ([15]crown-5 $\cdot H_3O)^+$ cation is less frequently observed, undoubtedly because the requisite threefold symmetry, complementary to that of the oxonium cation, is not present in the ligand [7]. Attempts to form the aza[18]crown-6 oxonium complex by this route have been unsuccessful as the preferred $(H_2aza[18]crown-6 \cdot H_2O)^+$ species is formed [8].

The structures of H_3O^+ and its higher homologues $(H_5O_2^+, H_7O_3^+, \text{etc.})$ have provided useful insights into mechanisms of proton transfer in aqueous solution. Alongside the impact of small protonated aqueous species there has been much interest in 'protonated water' clusters, particularly with regard to the most stable configuration for a protonated aqueous system. The most basic forms of the cation are those described by Eigen (H_3O^+) and Zundel $(H_5O_2^+)$ though neither have long-term stability in aqueous solution. In 2004 two reports [9,10] were published that gave good evidence, based on infrared signatures, of stable clusters with the composition $H^+(H_2O)_{21}$, as had been indicated by mass spectrometric data thirty years earlier. Based on O—H stretching frequencies it was shown that one- and two-dimensional networks collapsed into three-dimensional cages at that 'magic number' of 21 water molecules. Similar stoichiometries have been seen for methane hydrate clathrates with the composition $CH_4(H_2O)_{20}$ implying that aqueous clathrates form cages based on dodecahedral geometries with the guest, H_3O^+ in the case of 'protonated water', at the centre.

Preparation of a liquid clathrate

$(H_3O[18]crown-6)^+ \ HCl_2^- \cdot n$ *toluene* (39)

Reagents	**Equipment**
[18]Crown-6 (**15**)	Test tube
Hydrogen chloride gas [CORROSIVE]	Glass pipette
Toluene [FLAMMABLE]	NMR tubes
Potassium chloride	
Distilled water	
Deuterochloroform (CDCl₃)	
[TOXIC; CARCINOGENIC]	

If hydrogen chloride is to be generated *in situ* the following will also be required:

Concentrated sulphuric acid [CORROSIVE]	Pressure equalized dropping funnel
Ammonium chloride	Büchner flask

Note: all stages of this reaction, up to the preparation of the NMR sample, must be performed in a fume hood.

Charge a test tube with [18]crown-6 (**15**) (0.13 g, 0.5 mmol), potassium chloride (0.04 g, 0.5 mmol) and toluene (10 mL). Dissolve the crown ether by gently swirling the flask. Note that at this stage the potassium chloride remains solid to give a heterogeneous mixture. Attach the pipette to either a regulated cylinder of dry HCl gas or to a HCl generating system using non-corroding, flexible tubing.* Position the end of the pipette well below the surface of the toluene. Carefully open the valve on the gas cylinder, or slowly start dripping the sulphuric acid onto the ammonium chloride, until a regular stream of bubbles is observed in the solution. At this point, if the system is free from water, there will be no apparent change to the reaction mixture. If necessary, remove the pipette briefly, add distilled water (1 drop) and replace the pipette. As the salt dissolves, a small amount of a dense, second phase separates from the bulk solution. Once the formation of the lower phase appears to be complete, remove the gas supply and shut off the cylinder or, in the case of the generator, stop adding acid and allow the remaining gas to dissipate. The lower phase contains $(H_3O[18]crown-6)^+$ $HCl_2^- \cdot n$ toluene (**39**) and is best analysed by ^1H NMR. Dissolve a drop of the lower clathrate layer in deuterochloroform (0.75 mL) and obtain the ^1H spectrum from 0 to 12 ppm. The spectrum should consist of four groups of signals, as shown in the lower trace in Figure 5.4. The highly acidic protons from H_3O^+ appear as a sharp singlet at 9.1 ppm, aromatic toluene protons at 7.3 to 7.1 ppm, [18]crown-6 protons at 3.7 ppm and protons from the toluene methyl group at 2.35 ppm. Knowing that the toluene integration adds up to eight protons per solvent molecule, and that each [18]crown-6 molecule is associated with 24 protons, it is possible to determine the value of n for the clathrate produced. A typical experiment gives n as 14 or 15. A similar analysis can be performed on the upper liquid clathrate layer to show that it contains only toluene (the upper trace in Figure 5.4).

This method can be used to prepare a liquid clathrate from aza[18]crown-6 (**17**) using the same amounts of crown, salt and toluene. As the methylene protons are no longer equivalent (due to the substitution of NH for O) this region of the spectrum is slightly more complex; however, a broader acid proton peak appears at 10.3 to 10.5 ppm indicating H_3O^+ formation. Interestingly, this does not survive crystallization: analysis of bond lengths indicates that the acidic proton probably interacts as strongly with the crown amine as with the included water molecule.

*A 'hydrogen chloride generator' is easily prepared, as follows. Fit a stoppered, pressure-equalized dropping funnel half-filled with concentrated sulphuric acid (*ca.* 10 mL) to a Büchner flask containing ammonium chloride (*ca.* 10 g); dripping the acid onto the base will generate hydrogen chloride gas. To keep the evolution of gas at a steady rate it may be necessary to agitate the flask carefully or to add larger portions of the acid. Dissolve the sodium acetate formed in water and neutralize prior to disposal; dispose of any unused acid cautiously in accordance with local safety regulations.

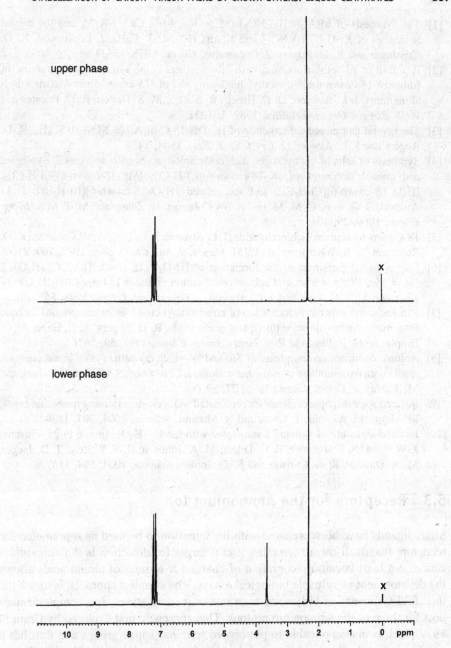

Figure 5.4 ^{1}H NMR of the upper and lower phases of $(H_3O[18]crown-6)^+$ HCl_2^- $\cdot n$ toluene

[1] The synthesis of $M[Al_2(CH_3)_6NO_3]$ ($M^+ = K^+$, Rb^+, Cs^+, NR_4^+), and the crystal structures of $K[Al_2(CH_3)_6NO_3]$ and $K[Al(CH_3)_3NO_3] \cdot C_6H_6$, J. L. Atwood, K. D. Crissinger and R. D. Rogers, *J. Organomet. Chem.*, 1978, **155**, 1.

[2] Reaction of trimethylaluminum with crown ethers – the synthesis and structure of (dibenzo-18-crown-6)bis(trimethylaluminum) and of (15-crown-5)tetrakis(trimethyl-aluminum), J. L. Atwood, D. C. Hrncir, R. Shakir, M. S. Dalton, R. D. Priester and R. D. Rogers, *Organometallics*, 1982, **1**, 1021.

[3] The crystal and molecular structure of $[K \cdot DB-18-C-6][AlMe_3NO_3] \cdot 0 \cdot 5C_6H_6$, R. D. Rogers and J. L. Atwood, *J. Cryst. Spec. Res.*, 1984, **14**, 1.

[4] Synthesis of salts of the hydrogen dichloride anion in aromatic solvents 2. Syntheses and crystal structures of $[K \cdot 18-crown-6][Cl-H-Cl]$, $[Mg \cdot 18-crown-6][Cl-H-Cl]_2$, $[H_3O \cdot 18-crown-6][Cl-H-Cl]$, and the related $[H_3O \cdot 18-crown-6][Br-H-Br]$, J. L. Atwood, S. G. Bott, C. M. Means, A. W. Coleman, H. Zhang and M. T. May, *Inorg. Chem.*, 1990, **29**, 467.

[5] 18-Crown-6 oxonium dichlorohydride, J. L. Atwood, S. G. Bott, A. W. Coleman, K. D. Robinson, S. B. Whetstone and C. M. Means, *J. Am. Chem. Soc.*, 1987, **109**, 8100.

[6] Use of metal carbonyls in the formation of $[H_5O_2^+ \cdot 15-crown-5][MOCl_4(H_2O)^-]$, (M = Mo, W), and a second sphere coordination complex in $[mer-CrCl_3(H_2O)(3) \cdot 15-crown-5]$, P. C. Junk and J. L. Atwood, *J. Organomet. Chem.*, 1998, **565**, 179.

[7] Synthesis and structural elucidation of novel uranyl crown-ether compounds isolated from nitric, hydrochloric, sulfuric, and acetic acids, R. D. Rogers, A. H. Bond, W. G. Hipple, A. N. Rollins and R. F. Henry, *Inorg. Chem.*, 1991, **30**, 2671.

[8] Anionic coordination complexes of Mo and W which crystallize from liquid clathrate media with oxonium ion–crown ether cations, J. L. Atwood, S. G. Bott, P. C. Junk and M. T. May, *J. Coord. Chem.*, 1996, **37**, 89.

[9] Infrared spectroscopic evidence for protonated water clusters forming nanoscale cages, M. Miyazaki, A. Fujii, T. Ebata and N. Mikami, *Science*, 2004, **304**, 1134.

[10] Infrared signature of structures associated with the $H^+(H_2O)_n$ ($n = 6$ to 27) clusters, J.-W. Shin, N. I. Hammer, E. G. Diken, M. A. Johnson, R. S. Walters, T. D. Jaeger, M. A. Duncan, R. A. Christie and K. D. Jordan, *Science*, 2004, **304**, 1137.

5.3 Receptors for the Ammonium Ion

Many ligands have been prepared with the intention to be used as supramolecular receptors for alkali metals. Another useful target for detection is the ammonium cation, not least because recognition of charged *N*-termini of amino acids allows the development of valuable biological assays. The simplest approach is to use the threefold symmetry of [18]crown-6, or one of its derivatives, as a complementary host for three of the ammonium protons. This approach, first explored by Cram in 1977 [1,2], is also applicable to protonated terminal amine groups and thus has a general utility. As the ion carries a delocalized positive charge the N—H bonds are more acidic than for the parent amine allowing strong hydrogen bonds to form with polyether oxygen atoms. Subsequent work by Sutherland illustrated how methylation of an azacrown ether could affect the preferred chirality of guests that had a terminal ammonium group [3].

The ability of a remote functional group to enhance cation binding by crown ethers was illustrated by Gokel who used *N*-pivot lariat ethers to bind the ammonium cation through a combination of crown and ethyleneoxy sidearm interactions [4]. Two ethyleneoxy units attached to the azacrown was found to give the highest stability constant for an aza[18]crown-6 complex with the ammonium cation. Too few donor atoms in the sidearm and no enhancement is observed: too many and the steric crowding, together with sidearm flexibility, also reduces the stability constant. This effect is shown in Figure 5.5, where simulations indicate weaker binding for lariat ethers with to few or too many donor atoms.

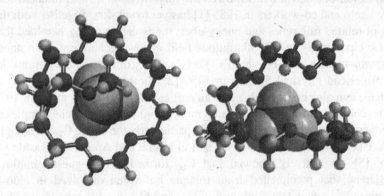

Figure 5.5 Ammonium--lariat ether complexes: top view (left) shows complementary tetrahedral symmetry with an ethoxymethyl ether derivative, side view (right) shows the slight destabilizing effect of a longer pendent ether

A logical extension to this strategy is to encapsulate the ammonium guest within a cryptand. Cram illustrated this with a series of 'hemispheraplex' complexes though, given the size of the central cavity, it is likely that the cation is not encapsulated but binds closely between two strands of the host molecules [5]. A better example is one in which a capped cyclotriveratrylene encapsulates an ammonium cation [6]. Needless to say these compounds will not be able to bind terminal groups on protonated natural and synthetic amines.

[1] Host–guest complexation 3. D. J. Cram, *J. Am. Chem. Soc.*, 1977, **99**, 6392.
[2] Host–guest complexation. 4. Remote substituent effects on macrocyclic polyether binding to metal and ammonium ions, S. S. Moore, T. L. Tarnowski, M. Newcome and D. J. Cram, *J. Am. Chem. Soc.*, 1977, **99**, 6398.
[3] The formation of complexes between aza derivatives of crown ethers and primary alkylammonium salts. Part 1. Monoaza derivatives, M. R. Johnson, I. O. Sutherland and R. F. Newton, *J. Chem. Soc., Perkin Trans. 1*, 1979, 357.
[4] 12-, 15-, and 18-Membered-ring nitrogen-pivot lariat ethers: syntheses, properties, and sodium and ammonium cation binding properties, R. A. Schultz, B. D. White, D. N. Dishong, K. A. Arnold and G. W. Gokel, *J. Am. Chem. Soc.*, 1985, **107**, 6659.

[5] Preorganisation – from solvents to spherands, D. J. Cram, *Angew. Chem. Int. Ed. Engl.*, 1986, **25**, 1039.

[6] Speleands – macropolycyclic receptor cages based on binding and shaping subunits – synthesis and properties of macrocycle-cyclotriveratrylene combinations, J. Canceill, A. Collet, J. Gabard, F. Kotzybahibert and J.-M. Lehn, *Helv. Chim. Acta*, 1982, **65**, 1894.

5.4   Purification of Fullerenes

The explosion of interest that followed the identification of buckminsterfullerene, C_{60}, by Kroto and co-workers in 1985 [1] has yet to subside. Together with the discovery of related fullerenes and buckytubes, the isolation of C_{60} heralded the start of a new chapter in the now ubiquitous field of nanotechnology. Like nanotechnology, which dates from Feynman's 'There's plenty of room at the bottom' lecture to the American Physical Society in 1959 [2], the identification of C_{60}'s 'soccer ball' shape can also be traced back to an earlier origin, Osawa's paper in 1970 [3].

In the early days of fullerene research most affordable commercial samples of C_{60} were contaminated with higher fullerenes, particularly the oblate C_{70}, making it hard to undertake C_{60}-specific chemistry. Then in March and April 1994 Raston [4] and Shinkai [5], respectively, showed that C_{60} formed a host–guest complex with calix[8]arene that precipitated from toluene but when dissolved in chloroform decomplexed to precipitate a mixture of C_{60} and C_{70} in a 9:1 ratio. Recrystallizing twice from toluene increased the purity of C_{60} to 99.5 per cent. Atwood later showed that cyclotriveratrylene could also bind C_{60} [6] and determined the structure of C_{60}-calixarene host–guest complexes [7]. Since then many examples of host–guest complexes containing C_{60} and other fullerenes have been reported, some of which are shown in Figure 5.6. Indeed the generality of this phenomenon has resulted in a patent [8]. The main factor in the purification mechanism seems to be the fit between the different orientations of the fullerene surface and the complexing agent. This explains why macrocycles with threefold symmetry, such as cyclotriveratrylene and oxacalix[3]arenes, as well as those with fivefold symmetry, the calix[5]arenes, are both effective complexants. Complexes involving the first group align the three macrocyclic aromatic rings with those of C_{60} that sit on a threefold symmetry axis: in the second group a fivefold symmetry axis aligns the macrocycle with C_{60}. Interestingly this inclusion behaviour even extends to the N-benzylazacalix[3]-arenes, such as **34**, even though it may be expected that the benzyl substituents are preferentially included into the calixarene cavity. That the inclusion phenomena takes days, rather than the hours observed for cyclotriveratrylene, indicates a degree of competition between C_{60} and the macrocycle's own benzyl substituents.

Recently it has been shown that single-walled carbon nanotubes and fullerenes are capable of blocking cellular transmembrane ion channels [9]. Given global interest in the potential toxicity of nanomaterials, including C_{60}, the macrocyclic

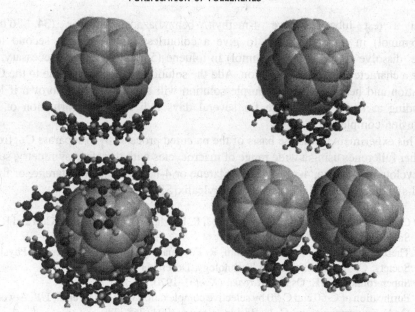

Figure 5.6 Supramolecular C_{60} complexes with bromooxacalix[3]arene (top left), cyclotrivera-trylene (top right), benzylcalix[5]arene (bottom left) and 4-t-butylcalix[6]arene (bottom right)

complexants discussed above should have the potential to form the basis of biomedical fullerene sensors. The experiment below illustrates how C_{60} can be complexed by a macrocycle with appropriately complementary symmetry elements. In the case of 4-methyl(N-benzyl)azacalix[3]arene, complexation is signalled by a gradual decrease in the solution's purple hue and concomitant precipitation of a brown solid. It is conceptually easy to extend this principle to the design of a C_{60}-binding macrocycle that forms complexes rapidly and signals their formation through an intense change in colour or fluorescence. Such a molecule would then fulfil the requirements of a C_{60} sensor.

Preparation of a C_{60}-macrocyclic complex

C_{60}-4-Methyl(N-benzyl)azacalix[3]arene (40)

Reagents
4-Methyl(N-benzyl)azacalix[3]arene (**34**)
Buckminsterfullerene (C_{60})
Toluene [FLAMMABLE]

Equipment
Test tubes

Note: this experiment should be carried out in a fume hood.

In a test tube, dissolve 4-methyl(N-benzyl)azacalix[3]arene (**34**, 36 mg, 0.05 mmol) in toluene (5 mL) to give a colourless solution. In a second test tube, dissolve C_{60} (7 mg, 0.01 mmol) in toluene (5 mL), heating if necessary, to give a characteristic purple solution. Add the solution of the macrocycle to the C_{60} solution and heat briefly. The purple solution will slowly turn pale brown if left standing at room temperature for several days indicating the formation of an inclusion complex of C_{60}.

This experiment forms the basis of the patented process [8] to separate C_{60} from higher fullerenes using a wide range of macrocycles with threefold symmetry, such as cyclotriveratrylene, 4-t-butylcalix[6]arene or 4-benzyloxacalix[3]arene, or five-fold symmetry, as in the case of 4-benzylcalix[5]arene.

[1] C_{60}: Buckminsterfullerene, H. W. Kroto, J. R. Heath, S. C. O'Brien, R. F. Curl and R. E. Smalley, *Nature*, 1985, **318**, 162.

[2] There's plenty of room at the bottom, R. Feynman, lecture to the American Physical Society, California Institute of Technology, December 29th 1959.

[3] Superaromaticity, E. Osawa, *Kagaku (Kyoto)*, 1970, **25**, 854.

[4] Purification of C-60 and C-70 by selective complexation with calixarenes, J. L. Atwood, G. A. Koutsantonis and C. L. Raston, *Nature*, 1994, **368**, 229.

[5] Very convenient and efficient purification method for fullerene (C_{60}) with 5,11,17, 23,29,35,41,47-octa-*tert*-butylcalix[8]arene-49,50,51,52,53,54,55,56-octol, T. Suzuki, K. Nakashima and S. Shinkai, *Chem. Lett.*, 1994, 699.

[6] Cyclotriveratrylene polarisation assisted aggregation of C_{60}, J. L. Atwood, M. J. Barnes, M. G. Gardiner, and C. L. Raston, *Chem. Commun.*, 1996, 1449.

[7] Symmetry-aligned supramolecular encapsulation of C_{60}: [$C_{60} \subset (L)_2$], L = p-benzyl-calix[5]arene or p-benzylhexahomooxacalix[3]arene, J. L. Atwood, L. J. Barbour, P. J. Nichols, C. L. Raston and C. A. Sandoval, *Chem. Eur. J.*, 1999, **5**, 990.

[8] Method for the purification and separation of fullerenes, J. L. Atwood and C. L. Raston, US patent 5711927, 1998.

[9] Single-walled carbon nanotubes are a new class of ion channel blockers, K. H. Park, M. Chhowalla, Z. Iqbal and F. Sesti, *J. Biol. Chem.*, 2003, **278**, 50212.

5.5 Making Molecular Boxes and Capsules

What if the size or reactivity of the guest precludes the use of a simple macrocycle as a host molecule? Large guests require large hosts, so the answer may be to use some form of directed ligand synthesis in which the guest is also the template. Similarly, encapsulation within a small cavity formed by two or more complementary species that interact through either covalent or hydrogen bonds may be the answer for reactive species. These chemical challenges are precisely the ones that should be addressed using supramolecular principles. Fortunately there are abundant examples of them in the literature.

Placing a guest molecule in a supramolecular cage, then subjecting it to conditions designed to initiate a particular reaction, has been shown to provide a route to chemistry that is otherwise hard to achieve. Cram cleverly used a hemicarcerand as a 'molecular test tube' to encapsulate cyclobutadiene by first including a lactam then irradiating the host-guest complex and generating the highly strained, and usually unstable, cyclobutadiene [1]. Heating the complex generated cyclooctatetraene that subsequently escaped its incarceration.

Using the well-known Crick–Watson and Hoogsteen hydrogen-bonding modes employed by DNA nucleotides, Gokel produced 'DNA boxes' in which diaza[18]-crown-6 was derivatized with adenine, thiamine, cytosine or a combination of two bases [2]. As expected, complementary nucleotide bases formed strong hydrogen bonds to produce dimers in the symmetrically substituted crowns: asymmetrically substituted crowns formed structures stabilized by intramolecular hydrogen bonds, as shown in Figure 5.7.

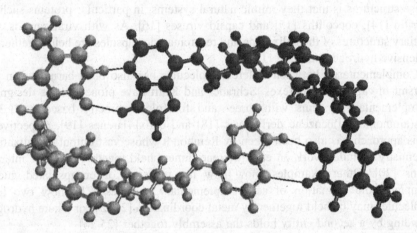

Figure 5.7 A molecular box formed by Crick--Watson interactions between functionalized azacrowns

In the late 1980s Rebek used Kemp's all-axial 1,3,5-triacid of cyclohexane as the basis for a number of compounds designed to bind substrates and, in some cases, catalyse reactions [3]. One of the three acid groups on Kemp's triacid was used to link two of the parent cyclohexane molecules together, using a variety of spacers, and the remaining groups were then presented as strong hydrogen-bonding sites. The effect that additional functionality in the spacer had on guest binding has been assessed computationally [4].

Container molecules of the types prepared by Gokel and Rebek, where dimers form primarily through hydrogen bonding and occasionally encapsulate guests, are relatively straightforward to conceptualize. Complementary arrays of binding sites are easy to design, if not always to synthesize [5,6]. Even compounds that do not seem to possess the necessary functionality to form dimers may exhibit unexpected behaviour under the right conditions. For example, the potassium hexafluorophosphate complexes of aza[15]crown-5 lariat ethers with cinnamyl, crotyl or allyl [7] sidearms all crystallize as dimers comprising two crowns, two perching cations and two bridging anions related through C_2 symmetry.

Symmetry is at the heart of the present endeavours to produce large supramolecular capsules. In 1997 the Atwood group reported a true supramolecule made up of six [4]resorcinarenes that encapsulated eight water molecules. The structure was highly symmetrical and conformed to one of the Archimedean solids [8]. The supramolecule revealed an important facet of calixarene behaviour, the ability to form crystalline solids that curved around templating species, rather than to form the commonly observed laminar structure comprising bilayer-like sheets. Once this concept had been grasped it was used to inform later syntheses which also produced 'spherical' supramolecular assemblies [9–13]. One crucial point about these structures is that they mimic natural systems, in particular proteins such as ferritin [14], coccoliths [15] and capsid viruses [16]. As with viral capsids, the tertiary structures of the calixarene and resorcinarene capsules are held together by extensive hydrogen bonding [17].

Complementarity between different molecules has also been harnessed in the pursuit of capsular complexes. Schrader and Kraft have pioneered the design of complementary systems with three- and fourfold symmetry based on 1,3,5-tris(aminomethyl)benzene derivatives [18] and calix[4]arenes [19], respectively. This approach has also been taken by Reinhoudt whose vast output of calixarene chemistry includes work on calix[4]arene dimers held together by ionic interactions [20]. Other examples show how the presence of encapsulated guests templates the formation of self-complementary capsules [21], how two host molecules may be held together by metal coordination [22–24] or where hydrogen bonding by a second entity holds the assembly together [25,26].

Perhaps nowhere has the use of symmetry at the nanoscale been more evident than in the elegant work of Seeman to use DNA itself as the vertices of Platonic solids [27]. In an interesting inversion the 'building block of life' has become the chemists' inanimate building block. Incorporation of specific sequences of complementary base pairs has made it possible to synthesize three-dimensional frameworks, though as yet these structures lack 'walls'. In a similar vein, one possible goal for future study is the manipulation of clathrins [28]. These proteins have a threefold, or triskelion, geometry and resemble three legs joined at the hip. Each leg contains a proximal section, knee, distal section, ankle and linker to a foot-like terminal protein domain. Each leg is about 475 Å (47.5 nm) long and twists around the joints. When triskelia interlock they form three-dimensional structures containing hexagonal and pentagonal openings that can be used to transport proteins across cell membranes. The flexibility of the triskelion framework allows different structures to form so that giant analogues of the C_{60}

framework can be made from 60 clathrin subunits and smaller supramolecules from 36 or 28 subunits.

[1] The taming of cyclobutadiene, D. J. Cram, M. E. Tanner and R. Thomas, *Angew. Chem. Int. Ed. Engl.*, 1991, **30**, 1024.

[2] Molecular boxes derived from crown ethers and nucleotide bases: probes for Hoogsteen vs Watson–Crick H-bonding and other base–base interactions in self-assembly processes, O. F. Schall and G. W. Gokel, *J. Am. Chem. Soc.*, 1994, **116**, 6089.

[3] Model studies in molecular recognition, J. Rebek, *Science*, 1987, **235**, 1478.

[4] Structure and binding for complexes of Rebek's acridine diacid with pyrazine, quinoxaline and pyridine from Monte Carlo simulations with an all-atom force field, E. M. Duffy and W. L. Jorgensen, *J. Am. Chem. Soc.*, 1994, **116**, 6337.

[5] Assembly and encapsulation with self-complementary molecules, J. Rebek, Jr, *Chem. Soc. Rev.*, 1996, 225.

[6] Self-assembling cavities: present and future, J. de Mendoza, *Chem. Eur. J.*, 1998, **4**, 1373.

[7] Formation of an organometallic coordination polymer from the reaction of silver(I) with a non-complementary lariat ether, P. D. Prince, P. J. Cragg and J. W. Steed, *Chem. Commun.*, 1999, 1179.

[8] A chiral spherical molecular assembly held together by 60 hydrogen bonds, L. R. MacGillivray and J. L. Atwood, *Nature*, 1997, **389**, 469.

[9] Structural classification and general principles for the design of spherical molecular hosts, L. R. MacGillivray and J. L. Atwood, *Angew. Chem. Int. Ed.*, 1999, **38**, 1019.

[10] Finite, spherical coordination networks that self-organize from 36 small components, M. Tominaga, K. Suzuki, M. Kawano, T. Kusukawa, T. Ozeki, S. Sakamoto, K. Yamaguchi and M. Fujita, *Angew. Chem. Int. Ed.*, 2004, **43**, 5621.

[11] Hydrogen-bonded molecular capsules are stable in polar media, J. L. Atwood, L. J. Barbour and A. Jerga, *Chem. Commun.*, 2001, 2376.

[12] Reversible encapsulation of multiple, neutral guests in hexameric resorcinarene hosts, A. Shivanyuk and J. Rebek, Jr, *Chem. Commun.*, 2001, 2424.

[13] A versatile six-component molecular capsule based on benign synthons – selective confinement of a heterogeneous molecular aggregate, G. W. V. Cave, M. J. Hardie, B. A. Roberts and C. L. Raston, *Eur. J. Org. Chem.*, 2001, 3227.

[14] Solving the structure of human H-ferritin by genetically engineering intermolecular crystal contacts, D. M. Lawson, P. J. Artymiuk, S. J. Yewdall, J. M. A. Smith, J. C. Livingstone, A. Treffry, A. Luzzago, S. Levi, P. Arosio, G. Cesareni, C. D. Thomas, W. V. Shaw and P. M. Harrison, *Nature*, 1991, **349**, 541.

[15] Crystal assembly and phylogenetic evolution in heterococcoliths, J. R. Young, J. M. Didymus, P. R. Bown, B. Prins and S. Mann, *Nature*, 1992, **356**, 516.

[16] Icosahedral virus particles as addressable nanoscale building blocks, Q. Wang, T. Lin, L. Tang, J. E. Johnson and M. G. Finn, *Angew. Chem. Int. Ed.*, 2002, **41**, 459.

[17] Toward mimicking viral geometry with metal-organic systems, J. L. Atwood, L. J. Barbour, S. J. Delgarno, M. J. Hardie, C. L. Raston and H. R. Webb, *J. Am. Chem. Soc.*, 2004, **126**, 13170.

[18] Self-assembly of ball-shaped molecular complexes in water, T. Grawe, T. Schrader, R. Zamard and A. Kraft, *J. Org. Chem.*, 2002, **67**, 3755.

[19] Capsule-like assemblies in polar solvents, R. Zamard, M. Junkers, T. Schrader, T. Grawe and A. Kraft, *J. Org. Chem.*, 2003, **68**, 6511.

[20] Guest encapsulation in a water-soluble molecular capsule based on ionic interactions, F. Corbellini, L. Di Costanzo, M. Crego-Calama, S. Geremia and D. N. Reinhoudt, *J. Am. Chem. Soc.*, 2003, **125**, 9946.

[21] Guest-induced capsular assembly of calix[5]arenes, D. Garozzo, G. Gattuso, F. H. Kohnke, P. Malvagna, A. Notti, S. Occhipinti, S. Pappalardo, M. F. Parisi and I. Pisagatti, *Tetrahedron Lett.*, 2002, **43**, 7663.

[22] Self-assembled nanoscale capsules between resorcin[4]arene derivatives and Pd(II) or Pt(II) complexes, S. J. Park and J.-I. Hong, *Chem. Commun.*, 2001, 1554.

[23] Inclusion of [60]fullerene in a homooxacalix[3]arene-based dimeric capsule cross-linked by a Pd(II)-pyridine interaction, A. Ikeda, M. Yoshimura, H. Udzu, C. Fukuhara and S. Shinkai, *J. Am. Chem. Soc.*, 1999, **121**, 4296.

[24] Complete selection of a self-assembling homo- or hetero-cavitand cage via metal coordination based on ligand tuning, K. Kobayashi, Y. Yamada, M. Yamanaka, Y. Sei and K. Yamaguchi, *J. Am. Chem. Soc.*, 2004, **126**, 13896.

[25] Self-assembly of cacerand-like dimers of calix[4]resorcinarene facilitated by hydrogen bonded solvent bridges, K. N. Rose, L. J. Barbour, G. W. Orr and J. L. Atwood, *Chem. Commun.*, 1998, 407.

[26] Molecular capsule constructed by multiple hydrogen bonds: self-assembly of cavitand tetracarboxylic acid with 2-aminopyramidine, K. Kobayashi, T. Shirasaka, K. Yamaguchi, S. Sakamoto, E. Horn and N. Furukawa, *Chem. Commun.*, 2000, 41.

[27] Nanotechnology and the double helix, N. C. Seeman, *Sci. Am.*, 2004, **290**, 34.

[28] Molecular model for a complete clathrin lattice from electron cryomicroscopy, A. Fotin, Y. Cheng, P. Sliz, N. Grigorieff, S. C. Harrison, T. Kirchhausen and T. Walz, *Nature*, 2004, **432**, 573.

5.6 Self-Complementary Species and Self-Replication

As well as the examples given in the previous section, in which complementarity of functional groups is used to form supramolecules, there are others in which self-complementarity generates interesting homodimeric complexes. A case in point is aminomethylbenzo[18]crown-6 which forms dimers when the amine sidearm is protonated [1]. The ammonium-[18]crown-6 motif is well known and, in this molecule, both host and guest are joined. More complex examples, such as those in Figure 5.8, include the self-complementary 'tennis balls' [2] and other cavitands [3–5] designed by Rebek who has shown how the relative positions and orientations of solvent encapsulated within dimeric assemblies can be determined, in the absence of X-ray data, by NMR. A concise treatment of the many approaches to supramolecular self-assembly are outlined in a 1995 review [6].

An obvious place to start when designing molecules that are intended to self-assemble is with existing biological examples. Nature has two prime methods for self-assembly: base pairing (as seen in RNA and DNA) and secondary polypeptide interactions. Both rely on hydrogen bonding and have proved an inspiration to supramolecular chemists. Consideration of the former led to Gokel's work on the

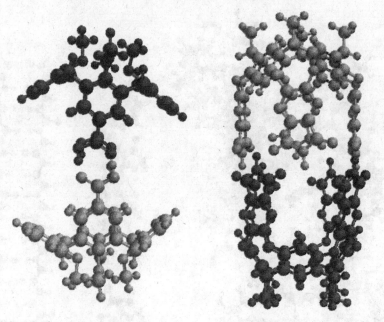

Figure 5.8 Capsule formation by self-complementary calixarenes (left) and cavitands (right)

incorporation of complementary base pairs into azacrown ethers, as discussed in the previous section, but many groups have been inspired by the propensity for amide linkages to self-associate, seen most clearly in the α-helix and β-sheet motifs found in proteins and illustrated in Figure 5.9. One of the most spectacular uses of this concept has been Ghaderi's preparation of cyclopeptides that spontaneously stack through extensive hydrogen bonding. The tubular supramolecules that form in this manner are able to puncture bacterial cell membranes, as the schematic in Figure 5.10 shows, and thus act as a novel class of antibiotics [7,8].

Perhaps the simplest way to induce self-complementary stacking in this way is to introduce two urea links into a cyclic structure and ensure that the torsional effects of the macrocycle induce the amide oxygens to adopt alternating geometries. This is exactly what was achieved by Shimizu in 2001. The two urea groups were incorporated within a 16-membered macrocycle that also contained two 1,3-xylene spacers and stacked, exactly as expected, in the crystalline state [9]. Since the original report, larger macrocycles based on the same principals have been synthesized [10,11]. An even simpler system, though perhaps of limited utility, is the co-crystallization of (1R,2R)-*trans*-1,2-diaminocyclohexane with and equimolar quantity of (1R,2R)-*trans*-cyclohexane-1,2-dicarboxylic acid [12]. If this particular combination of chiral cyclohexane derivatives spontaneously self assembles there must presumably be the potential for other molecules containing the same functional groups to form similar structures. Suitably derivatized cyclohexane-containing crown ethers would be good candidates for study in this regard.

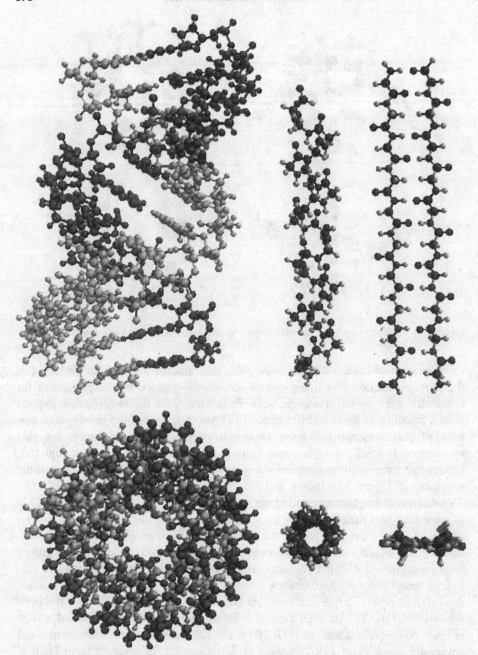

Figure 5.9 Biological hydrogen-bonding motifs: helical DNA (left), α-helical polyglycine (centre) and β-sheet polyglycine (right)

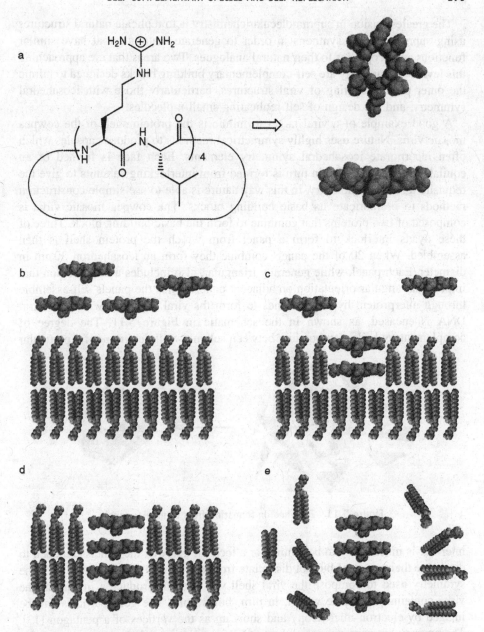

Figure 5.10 Self-assembling tubes of cyclic peptides and their effect on cell membranes: (a) cyclooctapeptide showing central cavity; (b) cyclic peptides approach a phospholipid membrane; (c) insertion initiated; (d) tube formation; (e) membrane disruption

The greatest vision in supramolecular chemistry is to replicate natural structures using supramolecular synthons in order to generate materials that have similar functions and properties to their natural analogues. Two areas that are approaching this level of complexity are self-complementary building blocks designed to mimic the outer protein coating of viral structures, particularly those with icosahedral symmetry, and the design of self-replicating small molecules.

A good example of a viral target to mimic is the protein shell of the cowpea mosaic virus. Nature uses highly symmetrical methods to produce capsules which often incorporate icosahedral symmetry elements. Each face is formed of an equilateral triangle which in turn is formed from interlocking subunits to give the requisite threefold symmetry. In this way nature is able to use simple construction methods to prefabricate its basic building blocks. The cowpea mosaic virus is composed of two proteins that combine to form the basic building block. Three of these dyads interlock to form a panel from which the protein shell is then assembled. When 20 of the panels combine they form an icosahedron 30 nm in diameter. Each panel, while generally triangular, also includes a hinge region that imposes an angular orientation on adjacent panels. Thus the panels self-assemble through interprotein hydrogen bonds to form the viral capsid in which the viral DNA is encased, as shown in the schematic in Figure 5.11. The degree of complementary hydrogen bonding between subunits will determine if a particular

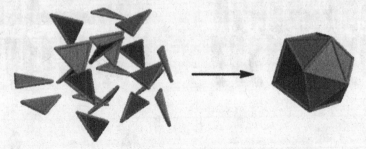

Figure 5.11   Icosahedron formation from triangular panels

interface is more likely to be a hinge or a lock. In this way the panels can open to discharge their contents but not dissociate from the viral capsid. Knowledge of the symmetry used to compose the viral shell was used to produce a mutant of the virus containing cysteine which, in turn, binds gold. The gold particles can be imaged by electron microscopy and show up as the vertices of a pentagon [13]. The analogy between viruses and artificial capsules is not lost on supramolecular chemists. If nanoscale supramolecular capsules can be engineered in different sizes and open under particular conditions then they will become an excellent method to dispense drugs where and when required.

The design of self-replicating systems has at its root the desire to understand, and perhaps compete with, nature's preferred method of information transfer,

DNA. The fundamental requirement of a self-replicating molecule is that it can be synthesized from two or more components for which the molecule itself is the template. It is also vital that the dimer that forms upon completion of any bond formation during the template-assisted synthesis dissociates readily so that more components can bind to the original template and its newly created twin. Lastly, there must be an available pool of components.

Three corollaries arise from this. The template is likely to have C_2 symmetry, to allow self-complementary dimers to form. The components will be attracted to the template molecule through complementary hydrogen bonding arrays or electro-static dipoles. The process of templating the components brings reactive sites in close proximity to each other: a general example of this is shown in Figure 5.12.

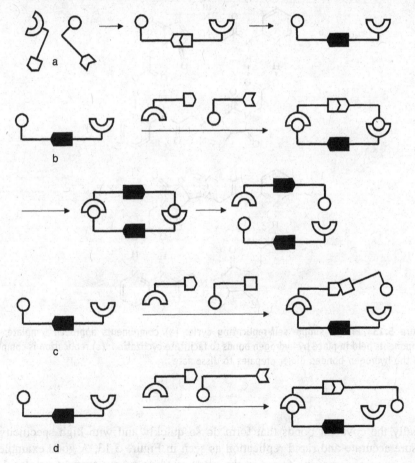

Figure 5.12   Aspects of simple self-replication: (a) two molecules form a covalent link; (b) the new molecule templates replication through complementary recognition sites; (c) mismatched functional groups fail to react; (d) mismatched recognition sites fail to template a replicate molecule

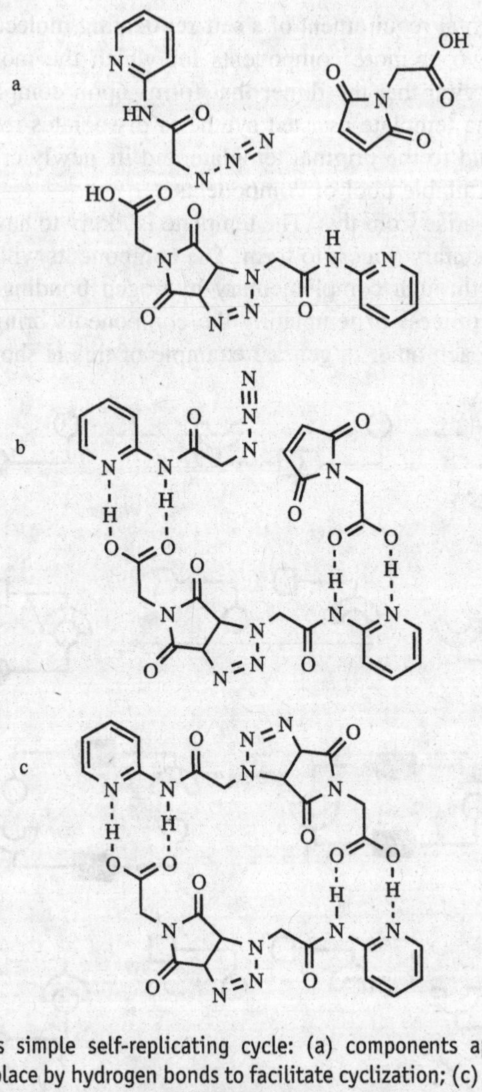

Figure 5.13 Philp's simple self-replicating cycle: (a) components approach template; (b) components held in place by hydrogen bonds to facilitate cyclization; (c) replication is complete and the hydrogen-bonded dimer prepares to dissociate

Ideally the covalent bonds that form do so quickly and with high specificity to ensure accurate and rapid replication as seen in Figure 5.13. A good example of these principles may be found in the work of Philp whose template molecule is complementary to an aminopyridine motif and a carboxylic acid [14]. The coupling reaction was between azide and maleimide groups and, notably, if the maleimide ester was substituted for the unprotected maleimide the rate of product formation was significantly lower. Other examples can be found in the work of Kiederowski, who has developed palindromic hexanucleotide sequences that replicate in the absence of enzymes [15], and the more conventional self-replicating systems of Rebek, whose 'replication cycle' [16] provoked a great deal of interest [17], and Ghaderi, who reported the first self-replicating abiotic peptide [18]. Good reviews of supramolecular self-replication have been published which focus on these fascinating systems [19–21].

[1] Stable supramolecular dimer of self-complementary benzo-18-crown-6 with a pendant protonated amino arm, O. P. Kryatova, S. V. Kryatov, R. J. Staples, E. V. Rybak-Akimoba, *Chem. Commun.*, 2002, 3014.

[2] Control of self-assembly and reversible encapsulation of xenon in a self-assembling dimer by acid–base chemistry, N. Branda, R. M. Grotzfeld, C. Valdés and J. Rebek, Jr, *J. Am. Chem. Soc.*, 1995, **117**, 85.

[3] Hydrogen-bonded capsules in polar, protic solvents, A. Shivanyuk and J. Rebek, Jr, *Chem. Commun.*, 2001, 2374.

[4] Assembly of resorcinarene capsules in wet solvents, A. Shivanyuk and J. Rebek, Jr, *J. Am. Chem. Soc.*, 2003, **125**, 3432.

[5] Hydrogen-bonded encapsulation complexes in protic solvents, T. Amaya and J. Rebek, Jr, *J. Am. Chem. Soc.*, 2004, **126**, 14149.

[6] Self-assembling supramolecular complexes, D. S. Lawrence, T. Jiang and M. Levett, *Chem. Rev.*, 1995, **95**, 2229.

[7] Cyclic peptides as molecular adaptors for a pore-forming protein, J. Sanchez-Quesada, M. R. Ghadiri, H. Bayley and O. Braha, *J. Am. Chem. Soc.*, 2000, **122**, 11757.

[8] Antibacterial agents based on the cyclic D,L-α-peptide architecture, S. Fernandez-Lopez, H.-S. Kim, E. C. Choi, M. Delgado, J. R. Granja, A. Khasanov, K. Kraehenbuehl, G. Long., D. A. Weinberger, K. M. Wilcoxen and M. R. Ghadiri, *Nature*, 2001, **412**, 452.

[9] Self-assembly of a bis-urea macrocycle into a columnar nanotube, L. S. Shimizu, M. D. Smith, A, D. Hughes and K. D. Shimizu, *Chem. Commun.*, 2001, 1592.

[10] Self-assembled nanotubes that reversibly bind acetic acid guests, L. S. Shimizu, A. D. Hughes, M. D. Smith, M. J. Davis, B. P. Zhang, H.-C. zur Loye and K. D. Shimizu, *J. Am. Chem. Soc.*, 2003, **125**, 14972.

[11] Self-assembly of dipeptidyl ureas: a new class of hydrogen-bonded molecular duplexes, T. Moriuchi, T. Tamura and T. Hirao, *J. Am. Chem. Soc.*, 2002, **124**, 9356.

[12] Supramolecular structures by recognition and self-assembly of complementary partners: an unprecedented ionic hydrogen-bonded triple-stranded helicate, P. Dapporto, P. Paoli and S. Roelens, *J. Org. Chem.*, 2001, **66**, 4930.

[13] Icosahedral virus particles as addressable nanoscale building blocks, Q. Wang, T. Lin, L. Tang, J. E. Johnson and M. G. Finn, *Angew. Chem. Int. Ed.*, 2002, **41**, 459.

[14] A structurally simple minimal self-replicating system, J. M. Quayle, A. M. Z. Slawin and D. Philp, *Tetrahedron Lett.*, 2002, **43**, 7229.

[15] Self-replication of complementary nucleotide-based oligomers, D. Sievers and G. Kiederowski, *Nature*, 1994, **369**, 221.

[16] Reciprocal template effects in a replication cycle, R. J. Pieters, I. Huc and J. Rebek, Jr, *Angew. Chem., Int. Ed. Engl.*, 1994, **33**, 1579.

[17] Kinetic analysis of the Rebek self-replicating system: is there a controversy? D. N. Reinhoudt, D. M. Rudkevich and F. de Jong, *J. Am. Chem. Soc.*, 1996, **118**, 6880.

[18] A self-replicating peptide, D. H. Lee, J. R. Granja, J. A. Martinez, K. Severin and M. R. Ghaderi, *Nature*, 1996, **382**, 525.

[19] The design of self-replicating molecules, M. M. Conn and J. Rebek, Jr., *Curr. Opin. Struct. Biol.*, 1994, **4**, 629.

[20] Autocatalytic networks: the transition from molecular self-replication to molecular ecosystems, D. H. Lee, K. Severin and M. R. Ghaderi, *Curr. Opin. Chem. Biol.*, 1997, **1**, 491.

[21] Minimal self-replicating systems, A. Robertson, A. J. Sinclair and D. Philp, *Chem. Soc. Rev.*, 2000, **29**, 141.

Appendix 1 Integrated Undergraduate Projects

Types of Projects

Most, if not all, of the examples of supramolecular synthons given here are amenable for use in undergraduate laboratories and may form the basis of further study. The analytical methods likewise are applicable to a wide variety of supramolecular systems and computational methods are limited only by the resources available to the students.

Two type of undergraduate laboratory exercise appear to be prevalent at present; the time-constrained 3-hour block that may be timetabled once a week, or more frequently, and the longer final year project which may be offered in either an intensive or extensive format. Experience has shown that even the simplest of experiments, such as Schiff base formation, is often unsuitable for a single laboratory session. Once the students have double-checked their laboratory scripts, assembled their reagents and equipment the session is often half over. Added to this is the time it takes for inexperienced students to recrystallize products, or the ever-present possibility that recrystallization fails to occur on demand, and a disastrous outcome is assured. Therefore, it is strongly suggested that any use of examples given in this book is restricted to longer projects or multiple laboratory sessions. Where a large number of students prepare the same compound, or a group of related compounds using the same techniques, it is advisable for the laboratory demonstrator to check the reaction and note any aspects that may cause problems. Unforeseen difficulties may range from limited equipment availability to variability in crystallization times. In addition it is vitally important to determine what the effects of scaling up the synthesis, from a single researcher to an entire laboratory group, will have on the levels of potentially hazardous by-products and other materials that will require disposal.

A Practical Guide to Supramolecular Chemistry   Peter J. Cragg
© 2005 John Wiley & Sons, Ltd

A Word on Project Safety

All students undertaking syntheses as part of their laboratory studies should be fully conversant with the safety data for each chemical used, each compound prepared and the correct method of disposal for all solvents, solutions and residues. The information in Appendix 2 is intended to be a guide in this respect: local safety regulations must always take precedence. Students should also follow standard operating procedures for any techniques described and discuss all the details of the experiments with a suitably qualified and experienced supervisor before starting an unfamiliar procedure.

Some Ideas

Several variations of the linear podands described in Chapter 1 can be envisaged. In the case of compound **1**, the most obvious changes involve using tri- and tetraethylene glycol in place of hexaethylene glycol to make shorter, more strained podands. Students could investigate their interactions with small alkali metals that ought to bind to the polyether regions of the podands, transition metals such as platinum or palladium that the nicotinyl termini should ligate, or a combination of the two. Alternatively other acid chlorides could be used to introduce termini with different properties. Would this yield compounds with the wealth of coordination motifs seen in the work of Hosseini? It would be a simple matter for a student to find out.

Many variations of compounds **4** have been reported, predominantly by Vögtle and co-workers, simply by substituting hydyoxybenzoic acids or similar aromatic alcohols for 8-hydroxyquinoline to give different terminal groups. Perhaps students could invent even more possibilities.

The original inspiration for compound **5** was Hannon's pyridylimine-terminated ligand. Clearly it is possible to use a variety of aromatic aldehydes to introduce the Schiff base link. In compound **6** salicylaldehyde was used to give convergent binding sites; however, 3- or 4-hydroxybenzaldehyde could be used to generate ligands with divergent sites. Would these compounds form supramolecular, hydrogen-bonded helical systems? How would they interact with transition metals that traditionally form complexes with Schiff bases? Finally, what would happen if these ligands were reduced? More flexible compounds will have far more conformational freedom. Will that generate new coordination motifs?

The tripodal compounds **6**, **7**, **8** and **9** are known to coordinate to lanthanides as well as anions; however, it would be interesting to see if they coordinate to transition metals too. There are examples of trigonal bipyramidal cobalt complexes in which one apical and three equatorial coordination sites are provided by a tripodal tetramine ligand with the other apical position being occupied by a counterion. Indeed, when a solution of compound **9** in chloroform is shaken with an aqueous solution of a cobalt salt an unusual purple solution forms. There are plenty of potential guests from the *d*-block that students could investigate.

The rigid tripodal compound **11** is just one of many that could be prepared from **10**. Other functionalized pyridines could be introduced to append a range of functional groups to the tripods. There is clearly the potential for the three aromatic substituents to form interdigitated dimers stabilized by π–π interactions. The formation of such species could be followed by NMR or maybe even UV-visible spectroscopy.

The compounds described in Chapter 2 are less amenable to synthetic variation. Many examples of porphyrin and phthalocyanine derivatives can be found in the literature, as well as their complexes with transition and main group metals. Likewise many combinations of 1,5-aromatic or pyridine dialdehyde derivatives and a diamine can be used to prepare cyclic Schiff bases though they often require a template. [18]Crown-6, **15**, can be used to illustrate liquid clathrate behaviour or act as phase transfer agents but cannot be derivatized further. Similar methods can, however, be used to synthesize crowns that contain aromatic groups or side chains. Azacrowns **16** and **17** can be derivatized in many ways. Two examples are given (compounds **18** and **19**) but many are known in the literature and ingenious students can undoubtedly devise ways of making many more. Many reactions of the azacrown lariat ethers can be devised, for example, students may wish to investigate the oxidizing effect of methanolic potassium permanganate on the allyl azacrowns. An alternative method of introducing *N*-substituted lariat ethers is to functionalize diethanolamine and react it with the appropriate polyethylene glycol ditosylate. This route is particularly attractive if large amounts of the lariat ether are required although a bottleneck may be encountered if Kugelrohr vacuum distillation is necessary to isolate the product and the apparatus only holds small volumes of the crude compound. Many azacrown lariat ethers can be purified by chromatography although they are often partially oxidized (indicated by a yellowing of compounds that should be colourless) by this process.

Only one cyclodextrin derivative is included, compound **20**, though it can be used to introduce pendent groups to the parent compound. Rotaxanes have been made by several research groups, notably those of Stoddart, Leigh, Sauvage and Gibson, though they are in general complex systems to prepare and even more complex to study. The Leigh group has also prepared a catenane in which two of the shuttles that form in the production of rotaxane **22** are interlocked. Their work, in particular, should be consulted by anyone interested in making catananes and other rotaxanes.

Whereas the examples in Chapter 1 are all essentially linear and those in Chapter 2 cyclic, the compounds described in Chapter 3 introduce a third dimension to supramolecular chemistry. Cyclotriveratrylenes, [4]resorcinarenes and pyrogallol[4]arenes, as exemplified by compounds **23**, **35** and **36**, are all prepared through the reaction of an aldehyde with an aromatic dialcohol, or derivative, in the presence of acid. Many variations on both the compounds and the reaction conditions used have been reported in the literature. Given the current interest in green chemistry, the solvent-free procedure of Raston and Atwood, by which cyclotriveratrylenes, [4]resorcinarenes, pyrogallol[4]arenes and even calix[4]arenes can be synthesized, may prove to be the most attractive method of all. Although it was not used to prepare the compounds described here, it is simple to reproduce and an efficient way to generate many different compounds in a short period of time. Undergraduates could use this method to produce a vast array of cyclic compounds from combinations of pyrogallol or resorcinarene with any number of aldehydes. Spectroscopic investigations could be undertaken to determine if the compounds self-assemble in solution.

The two calix[4]arenes, **24** and **26**, are examples of useful derivatives: the former binds lanthanides to form luminescent complexes, the latter, prepared from compound **25**, is water soluble and crystallizes in a variety of forms, from laminates to nanospheres. Other derivatives are to be found in the literature. Simple derivatives, such as the *O*-alkyl compounds, are used as a basis for upper rim functionalization: treatment of these compounds with nitric acid replaces the upper *t*-butyl groups with nitrates. Nitrocalixarenes can be reduced to the corresponding amines to generate a platform for further extension of the cavity and have been used to bind metals or small molecules. Oxa- and azacalix[3]arenes represent crown ether–calixarene hybrids and have binding modes reminiscent of both classes of molecules. Many alternative upper rim substituents can be introduced and, in the case of the azacalixarenes, the *N*-substituent can easily be varied.

Cryptand, **37**, and its reduced form, **38**, come from the extensive work of Nelson and McKee. Other cryptands incorporating aromatic spacers are also described in their papers and can be made in a similar fashion. The compounds bind cations, specifically transition metals, and anions. As such they perhaps represent the ultimate three-dimensional supramolecular systems as the compounds encrypt isolated ion pairs.

All of the above compounds have the potential for *in silico* molecular modelling to probe their likely affinities for guests or their potential to signal when guest binding occurs. Many of them can be studied spectroscopically to detect the formation of host-guest complexes and to determine stoichiometries. Students may want to investigate not only standard NMR methods to follow guest binding but also two-dimensional methods, which may indicate if the guests bind symmetrically within a preorganized cavity.

An almost infinite combination of the compounds described above and molecular or ionic guests can be considered for investigation by interested and well-motivated undergraduates. The investigations may require the students to prepare the molecules themselves or to investigate supramolecular phenomena using compounds prepared by the laboratory instructor. Either way, the vast scope of supramolecular chemistry should continue to inspire future generations of scientists.

Appendix 2　Reagents and Solvents

The following descriptions should be read in conjunction with any specific instructions given for the relevant experiment(s) in the text.

Hazards	Any particular hazards as recorded by COSHH/OSHA
Precautions	Standard (lab coat, safety glasses, gloves, work where possible in a fume hood)
	Intermediate (as above but reagent should only be used in a fume hood)
	Advanced (as above but see specific experiments for guidance)
Effects	Route and effects of acute exposure (I, inhalation; S, skin absorption; E, eyes; NV, nausea and/or vomiting; D, dizziness; L, lachrymatory effects; R, respiratory problems)
Incompatibles	Any specific known incompatible effects, e.g. violent reaction with water, other than those listed under Hazards
Disposal	General guidelines for initial disposal (D, dilute with water; neutralize, if necessary, with an appropriate inorganic acid or base and flush down drain; C, chlorinated solvent disposal; NC, non-chlorinated solvent disposal; S, solid disposal in accredited landfill site or burned in a chemical incinerator; R, refer to local safety officer)
Notes	Any specific comments.

A Practical Guide to Supramolecular Chemistry　Peter J. Cragg
© 2005 John Wiley & Sons, Ltd

Appendix 2 Reagents and Solvents

	Hazards	Precautions	Effects	Incompatibles	Disposal	Notes
Acetaldehyde	Flammable Lachrymator	Intermediate	I, S, E; R		S	May form peroxides
Acetic acid, glacial	Corrosive	Standard	I, S, E	Bases	D	
Acetone	Flammable	Standard	I, E, D		NC	
Acetonitrile	Flammable	Standard	I, S		NC	
Acetonitrile-d₃	Possible teratogen Flammable	Standard	I, S		R	
Allyl bromide	Possible teratogen Toxic Flammable Lachrymator	Intermediate	I, S, E, D	Reacts with water	S	
N-Allylaza[15]crown-5	UNKNOWN	Standard	UNKNOWN		S	
Aluminium trichloride	Corrosive	Advanced	I, S, E	Reacts with water to liberate corrosive and flammable gases	R	Avoid breathing dust
Ammonium chloride	Irritant	Standard	I, S, E		S	
Aza[15]crown-5	UNKNOWN	Standard	UNKNOWN		S	
Aza[18]crown-6	UNKNOWN	Standard	UNKNOWN		S	
Benzaldehyde	Irritant Toxic	Standard	UNKNOWN		S	
Benzylamine	Corrosive	Intermediate	I, S, E, R		S	
N-Benzyl-4-methyl-triazacalix[3]arene	UNKNOWN	Standard	UNKNOWN		S	
2,6-Bis(hydroxymethyl)-4-t-butylphenol	UNKNOWN	Standard	UNKNOWN		S	
2,6-Bis(hydroxymethyl)-4-methylphenol	UNKNOWN	Standard	UNKNOWN		S	
2,6-Bis(hydroxymethyl)-4-phenylphenol	UNKNOWN	Standard	UNKNOWN		S	
1,19-Bis(isonicotinyloxy)-4,7,10,13,16-pentaoxaheptadecane	UNKNOWN	Standard	UNKNOWN		S	
1,9-Bis(8-quinolinyloxy)-3,6-dioxanonane	UNKNOWN	Standard	UNKNOWN		S	
Buckminsterfullerene	Irritant	Standard	I, S, E, R		S	
t-Butanol	Flammable	Standard	I, E		NC	
t-Butylcalix[4]arene	UNKNOWN	Standard	UNKNOWN		S	

Compound	Hazard	Grade	Codes	Storage	Notes
t-Butylcalix[4]arenetetraacetamide	UNKNOWN	Standard	UNKNOWN	S	
4-t-Butyloxacalix[3]arene	UNKNOWN	Standard	UNKNOWN	S	
4-t-Butyloxacalix[3]arene-tris(diethylacetamide)	UNKNOWN	Standard	UNKNOWN	S	
4-t-Butyloxacalix[3]arene-tri(acetic acid)	UNKNOWN	Standard	UNKNOWN	S	
4-t-Butylphenol	Irritant	Standard	I, R	S	Inhalation may be fatal
Calix[4]arene	UNKNOWN	Standard	UNKNOWN	S	
Chloroform	Toxic Carcinogen	Intermediate	I, E, NV, D	C	
Chloroform-d	Carcinogen Toxic	Intermediate	I, E, NV, D	R	
N-Cinnamylaza[18]crown-6	UNKNOWN	Standard	UNKNOWN	S	
Cinnamyl bromide	Toxic	Standard	I, E	S	
Copper (II) bromide	Irritant	Standard	S	S	
Copper phthalocyanine		Standard		S	Intense pigment can stain
p-Cresol (4-methylphenol)	Poison	Standard	I, S, R	S	May be fatal if ingested
[18]Crown-6	Irritant	Standard	S	S	
β-Cyclodextrin hydrate	UNKNOWN	Standard	UNKNOWN	S	
β-Cyclodextrin tosylate	UNKNOWN	Standard	UNKNOWN	S	
Cyclotriveratrylene		Standard		NC	
Deuterium oxide	Irritant	Standard	I, E, L	S	
1,5-Diazabicyclo[4.3.0]non-5-ene	Toxic	Standard	I, E	C	
Dichloromethane	Harmful	Intermediate	I, E	S	
Diethanolamine	Toxic	Intermediate	I, S, E, R	S	
N,N-Diethylchloroacetamide	Lachrymator	Intermediate	I, S, E, R	S	
Diethylene glycol	Teratogen	Intermediate	I, S, E, R	S	
Diethyl ether	Flammable	Standard	I, D	NC	
Dimethylformamide	Harmful	Standard	I, S, E	NC	
Dimethyl sulfoxide	Irritant	Standard	I, S, E	NC	
Dimethyl sulfoxide-d6	Irritant	Standard	I, S, E	R	
Dimethyl 2,6-pyridinedicarboxylate	Irritant	Standard	I, S, E, R	S	

(Continues overleaf)

	Hazards	Precautions	Effects	Incompatibles	Disposal	Notes
Diphenylethylamine	Irritant	Standard	I, S, E, R		S	
1,4-Dioxane	Flammable	Standard	E, D		NC	May form explosive peroxides
	Carcinogen					
Distilled water		Standard			D	
Ethanol	Flammable	Standard	I, D, NV		NC	
	Toxic					
Ethyl acetate	Flammable	Standard	I, D		NC	
Ethylenediamine	Flammable	Intermediate	I, R		S	
Formaldehyde (37% aq.)	Toxic	Intermediate	I, S, E, R		S	May cause inheritable genetic damage
	Mutagen					
	Carcinogen					
Fumaryl bisamide	UNKNOWN	Standard	UNKNOWN		S	
Fumaryl chloride	Harmful	Standard	I, S, E, R		S	
C-Heptylpyrogallol[4]arene	UNKNOWN	Standard	UNKNOWN		S	
Hexaethylene glycol	Irritant	Standard	I, S, E, NV		S	
Hexane	Flammable	Standard	I, S, E, D		NC	May cause impaired fertility
	Possible neurotoxin					
Hydrobromic acid in acetic acid	Corrosive	Standard	I, S, E, R	Reactions often generate heat and/or harmful gases	D	
	Oxidant					
Hydrochloric acid	Corrosive	Standard	I, S, E, R	Reactions often generate heat and/or harmful gases	D	
	Oxidant					
Hydrogen chloride gas	Corrosive	Advanced	I, S, E, R		R	May be fatal if inhaled
	Oxidant					
3-Hydroxybenzaldehyde	Irritant	Standard	I, S, E, R		S	
8-Hydroxyquinoline	Possible carcinogen	Standard	I, S, E		S	
Isonicotinoyl chloride	Corrosive	Intermediate	I, S, E, R, L		S	
Isophthalaldehyde	UNKNOWN	Standard			S	
Isophthaloyl chloride	Corrosive	Intermediate	I, S, E, L		S	

Compound	Hazard	Level	Codes	Notes	Disposal
Magnesium sulphate	Irritant	Standard	I		S
Methanol	Flammable, Toxic	Standard	I, S, E, D		NC
Methanol-d_4	Flammable	Standard	I, E, D		R
2-Methoxyethyl ether (diglyme)		Standard	I, S, E, R		NC
N,N''-(4,4'-Methylenebiphenyl)bis(salicylideneimine)	UNKNOWN	Standard	UNKNOWN		S
4,4-Methylenedianiline	Harmful, Possible carcinogen	Standard			S
N-Methylquinuclidinium iodide	UNKNOWN	Standard	UNKNOWN		S
C-Methyl[4]resorcinarene	UNKNOWN	Standard	UNKNOWN		S
Nitric acid	Corrosive, Oxidant	Standard	I, S, E	Reactions often generate heat and/or harmful gases	D
Octanal	Flammable	Flamable	S, E, R	Store below 5 °C	S
Paraformaldehyde	Toxic, Mutagen, Carcinogen	Intermediate	I, S, E, R	May cause inheritable genetic damage	S
Phenol	Corrosive, Toxic	Standard	I, S, E, R		S
4-Phenylphenol	Irritant	Standard	I, S, E		S
4-Phenylpyridine	Irritant	Standard	I, S, E		S
Phthalonitrile	Irritant	Standard	I, S, E		S
Potassium borohydride	Toxic	Advanced	I, S, E, R	Reacts with water to liberate harmful gases	R
Potassium t-butoxide	Flammable solid, Corrosive	Advanced	I, S; E; R	Reacts with water to liberate harmful gases	R
Potassium carbonate	Irritant	Standard	I, S, E		S
Propanoic acid	Corrosive	Intermediate	I, S, E, R		S
Propan-2-ol	Flammable, Harmful	Standard	I, S, E, R		S
Pyrogallol	Harmful	Standard	I, S, E, R		S

(Continues over leaf)

	Hazards	Precautions	Effects	Incompatibles	Disposal	Notes
Pyrrole	Harmful	Intermediate	I, S, E, NV		S	Ingestion may be fatal
Quinuclidine	UNKNOWN	Standard	UNKNOWN		S	
Reduced *tren*-cryptand	UNKNOWN	Standard	UNKNOWN		S	
Resorcinol	Harmful	Standard	I, S, E, NV, R		S	
Rotaxane	UNKNOWN	Standard	UNKNOWN		S	
Salicylaldehyde	Irritant	Standard	I, S, E		S	
Silica	Irritant	Standard	I, S, E, R		S	
Sodium *t*-butoxide	Flammable solid Corrosive	Advanced	I, S, E, R	Reacts with water to liberate harmful gases	R	
Sodium carbonate, aqueous		Standard			D	
Sodium chloride, aqueous		Standard			D	
Sodium hydride (60% oil suspension)	Flammable solid	Advanced	I, S, E, R	Reacts with water to liberate harmful gases	R	
Sodium hydroxide	Corrosive	Standard	S, E, R		D	
Sodium sulphate	Irritant	Standard	I, S, E		S	
Sulphonatocalix[4]arene	UNKNOWN	Standard	UNKNOWN		S	
Sulphuric acid	Corrosive Oxidant	Standard	I, S, E, R	Reactions often generate heat and/or harmful gases	D	
Tetraethylene glycol	Irritant	Standard	I, S, E, NV		S	
Tetraethylene glycol ditosylate	Irritant	Standard	I, S, E		S	
Tetrahydrofuran	Flammable Irritant	Standard	I, S, E, NV	May form explosive peroxides	NC	May have narcotic effects
Tetralactam	UNKNOWN	Standard	UNKNOWN		S	
Tetraphenylporphyrin	Irritant	Standard	S, E		S	
Toluene	Flammable Toxic	Standard	I, S, D		NC	
4-Toluenesulphonic acid	Corrosive	Standard	I, S, E, R		S	
4-Toluenesulphonic anhydride	Corrosive	Standard	I, S, E, R		S	
4-Toluenesulphonyl chloride	Corrosive	Standard	I, S, E, R		S	

Name	Hazard		Codes	
Tren-cryptand	UNKNOWN	Standard	UNKNOWN	S
1,3,5-Tribromomethyl-2,4,6-trimethylbenzene	UNKNOWN	Standard	UNKNOWN	S
Triethylamine	Flammable Corrosive	Advanced	I, S, E, L, R	S
Triethylene glycol	Irritant	Standard	I, S, E, NV	S
Triethylene glycol ditosylate	Irritant	Standard	I, S, E	S
1,3,5-Trimethylbenzene (mesitylene)	Irritant	Standard	I, S, E	NC
Phenyloxacalix[3]arene	UNKNOWN	Standard	UNKNOWN	S
1,3,5-Tri(4-phenylpyridyl)-2,4,6-trimethylbenzene	UNKNOWN	Standard	UNKNOWN	S
Tris(2-aminoethyl)amine (*tren*)	Toxic	Standard	I, S, E, R	S
N^1,N^1,N^1-Tris(2-aminoethylamino)methylbenzene)	UNKNOWN	Standard	UNKNOWN	S
N^1,N^1,N^1-Tris(2-(2-aminoethylamino)methyl)phenol	UNKNOWN	Standard	UNKNOWN	S
N^1,N^1,N^1-Tris(2-(2-aminoethylimino)methyl)phenol	UNKNOWN	Standard	UNKNOWN	S
N^1,N^1,N^1-Tris(3-(2-aminoethylimino)methyl)phenol	UNKNOWN	Standard	UNKNOWN	S
Tris(2-aminoethyl)amine	Harmful	Standard	I, S, E, R	S
Veratrole	Irritant	Standard	I, S, E	S
o-Xylene	Flammable	Standard	I, S, E, D	NC
p-Xylenediamine	Corrosive	Standard	I, S, E	S

Index